Praise for Carlin Flora's

Friendfluence

"Awash in arresting insights with practical implications, many of them counterintuitive." —*The Huffington Post*

"Intriguing. . . . A convincing case for nurturing friendships in many of the same ways we nurture relationships with partners and other family—both online and off."

—*Kirkus Reviews*

"[Flora's] interdisciplinary discussion draws on scientific research, philosophy and anecdotes to examine friendship across a lifespan, from playground pals to adolescent and adult relationships. . . . Compelling. . . . Discloses many of friendship's secrets." —*Publishers Weekly*

"*Friendfluence* is so persuasive that the minute I put the book down, I made three dates to see friends."

—Gretchen Rubin, *New York Times* bestselling author of *The Happiness Project*

"Combining the latest research with engaging stories . . . Carlin Flora has written a delightful book on the power of friendship." —James H. Fowler, coauthor of *Connected* and professor of medical genetics and political science at the University of California, San Diego

"A seminal contribution to the literature on friendship. In this meticulously researched and eminently readable book, journalist Carlin Flora has mined the extant research on this complex topic and woven it together with real-life examples (both her own and others). In so doing, she helps explain how these relationships evolve and their impact on our day-to-day lives from childhood through adulthood."

—Irene Levine, Ph.D., author of *Best Friends Forever: Surviving a Breakup with Your Best Friend*

"Happiness and success begin with self-knowledge, and as Carlin Flora shows us in her compelling and delightful book, *Friendfluence*, the key to understanding yourself may well lie in your friendships, past and present. This is a must-read for anyone looking to experience greater well-being. . . . In other words, for everyone."

—Heidi Grant Halvorson, Ph.D., author of *Succeed* and associate director of the Motivation Science Center, Columbia University Business School

"A captivating read about an eternally fascinating subject—friendship. Flora's easy-to-read prose blends narrative and scientific research seamlessly. You will finish the book with a better understanding of why good friends are worth keeping."

—Jane Gradwohl Nash, professor of psychology and one of the "girls from Ames"

"[A] lively account of both the science and poetry of friendship."

—Dalton Conley, Ph.D., author of *The Pecking Order* and professor of sociology at New York University

Carlin Flora

Friendfluence

Carlin Flora was on the staff of *Psychology Today* for eight years, most recently as features editor. She is a graduate of the University of Michigan and Columbia University School of Journalism and has written for *Discover*, *Glamour*, *Women's Health*, and *Men's Health*, among others. She has also appeared on *The Oprah Winfrey Show*, *The Today Show*, CNN, Fox News, and *20/20*. She lives in Queens, New York.

www.carlinflora.com

Friendfluence

THE SURPRISING WAYS FRIENDS MAKE US WHO WE ARE

Carlin Flora

ANCHOR BOOKS
A DIVISION OF RANDOM HOUSE LLC
NEW YORK

FIRST ANCHOR BOOKS EDITION, OCTOBER 2013

 Published in the United States by Anchor Books, a division of Random House LLC, a Penguin Random House Company, New York, and in Canada by Random House of Canada Limited, Toronto. Originally published in hardcover in the United States by Doubleday, a division of Random House LLC, New York, in 2013.

Anchor Books and colophon are registered trademarks of Random House LLC.

The Library of Congress has cataloged the Doubleday edition as follows:
Flora, Carlin.
Friendfluence : the surprising ways friends make us who we are / Carlin Flora.
p. cm.
Includes bibliographical references and index.
1. Friendship. I. Title.
BF575.F66F56 2013
158.2'5—dc23
2012019366

Anchor ISBN: 978-0-307-94695-9

www.anchorbooks.com

Printed in the United States of America

10 9 8 7 6 5 4 3 2 1

To Mom and Dad

CONTENTS

Friendfluence

INTRODUCTION

Each friend represents a world in us, a world not possibly born until they arrive.
—ANAÏS NIN

WHEN I WAS FIFTEEN, MY FAMILY MOVED from North Carolina to Michigan. The relocation was difficult for one reason above all: I had to leave behind my friends. For the first few months at my new school I was a puddle of tears as I attempted to connect to other kids but didn't feel I could truly be myself. I read and reread letters from my old friends and felt painfully excluded from their latest escapades. Then one day I saw *them* up in the bleachers during a pep rally: They were a boisterous group of "alternative" girls (this was the '90s) who were nonetheless not *too* alternative, I soon learned: They were adventurous and artsy but still cared about getting good grades. From the first time I sat at their lunch table, my isolation began to subside. I started to feel excited about life again.

I was sentimental to begin with, which is probably why leaving my North Carolina friends was so painful. But my experience is far from unique: Friendship is a crucial facet of life, and not just for melodramatic teenage girls.

During the eight years I worked at *Psychology Today* magazine as a writer and editor, I noticed a steady increase in scientific findings about friendship. Study after study pointed to its surprising benefits. Who knew that friendship could be so good not only for one's mood but for one's health? Solid friendships can help you shed pounds, sleep better, stop smoking, and even survive a major illness. They can also improve memory and problem-solving abilities, break down prejudices and ethnic rivalries, motivate people to achieve career dreams, and even repair a broken heart. Yet very few of the many social science and self-help books that crossed my desk covered all of these aspects of friendship. Walk through the relationships section of any bookstore and you will be overwhelmed with titles about finding and keeping a romantic partner or parenting a child. An alien perusing this body of literature might assume that lovers and families are the only relationships we humans have.

Of course we also have friends. We might think all of our traits and life decisions can be traced back to our genes or the influence of our parents or partners, but it has become increasingly clear that our peers are stealth sculptors of everything from our basic linguistic habits to our highest aspirations. And while friendships are a staple in most of our lives, very few of us are fully aware of the effect friends have on our personal growth and happiness.

The converse holds true, too: A person without friends will become unhappy or worse. Loneliness sends the body and mind into a downward spiral. A lack of friends can be deadly.

EVOLUTIONARY PSYCHOLOGISTS THEORIZE that friendship has roots in our early dependence on others for survival. Having a friend help you hunt, for instance, made it more likely that you and your family—and your hunting buddy and his family—would have food cooking over the fire. While most of us no longer rely on friends for house building or meal gathering, we still have a strong need for them. Anthropologists have found compelling evidence of friendship throughout history and across cultures. Universally, we're built to care deeply about select people outside of our kin group. It's hard to construct a personal life history that doesn't include important parts for one's friends.

Now happens to be a prime time for increasing our awareness of how friends affect us. Friends are not just more important than you might think; they actually are becoming more important sociologically. In his 2004 book *Urban Tribes*, journalist Ethan Watters posed the question: "Are friends the new family?" Watters entertainingly depicted city-dwelling buddies who relied on one another throughout their twenties and even thirties, as they delayed marriage and found their vocational callings—a phenomenon of his class and age-group. While big, stable "tribes" might not characterize most Americans' social circles, people of all ages (and from all areas of the country) are relying on friends to fulfill duties traditionally carried out by blood relatives or spouses.

The median age of first marriage is still rising: In 2010 it was 28.7 for men and 26.5 for women, up from 27.5 and 25.9 in 2006. Americans aren't merely delaying marriage; many are divorced or widowed or are opting out completely. One hundred million or so Americans (that's almost half of all adults) are not married, and a 2006 Pew Research study found that 55 percent of singles are not looking *ever* to get married.

College students and young adults seem to be less inclined to have steady romantic relationships and are instead "hooking up" casually with one another. It stands to reason that without the psychological support of a serious boyfriend or girlfriend, this group is also relying on friends more than their demographic equivalents have in the past.

Sociologist Eric Klinenberg points out that "more people live alone now than at any other time in history." In cities such as Atlanta, Denver, Seattle, San Francisco, and Minneapolis, at least 40 percent of all households are made up of a single person. Klinenberg blows apart the stereotype of the lonely, quirky singleton by concluding that these people, whether young or elderly, socialize with friends more than do those who live with partners and families.

So, for the increasing number of people who are not living in traditional family structures, friends are often primary ties, providing close emotional support and "instrumental" help as well. It's not necessarily an either/or proposition, where friends must replace family, however. Singles are often close to their parents, nieces and nephews, and siblings, after all. But friends, in part because they are free of the heavy weight of obligation, can be even more beneficial and life-enhancing than relatives, particularly if they live near us.

It's not just single people for whom friends matter—a lot. Friends are also important for parents and those who are married or living with a romantic partner. Time with friends is actually our most pleasant time: We are most likely to experience positive feelings and least likely to experience negative ones when we are with friends compared to when we are with a spouse, child, coworker, relative, or anyone else. We're not surprised when we hear people grumbling about how they have to attend a family holiday party, yet it would puzzle us to hear the same people complain about having to go to a celebration full of their friends.

Why do we prefer spending time with our friends over our families? Some say it is because we pick our friends (God's consolation prize) while we don't pick our families. Insofar as we choose our spouses and decide to have children, we do have some say over our families. More likely, our time with our pals is more enjoyable because of our expectations. When we're with friends, we bring sympathy and understanding and leave out some of the grievances we carry into interactions with family members. We tend to demand less from friends than we do from relatives or our romantic partners, and each friend provides us distinct benefits. For instance, one might be our confidante, another might make us laugh, while a third is our go-to person for political discussion. We don't insist that they be everything to us; thus we are less disappointed when a friend falls short in a certain way than we are when a parent or spouse does the same.

When working parents devote every scrap of free time to their children, their friendships are the first thing to slide. We know from research (and our own intuition quickly confirms this) that expecting one's spouse to be everything is a recipe for disaster. Leaning on friends for intellectual stimulation, emotional support, and even just fun activities relieves the pressure of the overheated nuclear family. Busy moms and dads would do well to stop considering friends to be a nonessential luxury.

Kids themselves might also be more friend centered than they were, say, fifty years ago. Back then only children made up 10 percent of American kids under the age of eighteen. The latest census reveals that the ratio of "onlies" has doubled. There are about fourteen million of them, and they are likely seeking out pals more because in-house playmates aren't available.

In some ways we put friendship up on a pedestal. Think of all the popular movies and TV shows (such as, um, *Friends*) about tight clans whose members see one another through life's awkward moments and dramatic trials alike. But if we understood how beneficial real friends are, I think we'd be less passive and more careful about how we treat them, even if other people, such as our partners or kids, officially occupy the primary places in our hearts.

Friendfluence, then, is the powerful and often unappreciated role that friends—past and present—play in determining our sense of self and the direction of our lives. In the pages ahead, you'll learn how friends affect us during different developmental phases. As children, we're attached to our parents but preoccupied with our pals. Preschoolers who have trouble making friends tend to go on to have bad relationships with younger siblings, for instance. As middle schoolers, kids who don't care what friends think of them do worse academically and socially in high school—and beyond. It's not just that good friends are nice to have; the skills one needs to make good friends are the very abilities one generally needs to be successful in life. ("Tiger Moms" should rethink sleepover bans if they want their children to thrive in the social jungle, for which there is no adequate cramming course.)

When we are teenagers, friends co-create our fledging identities. Drug use, smoking, and early sexual activity are highly influenced by peer behaviors as well as parental behaviors. The often overlooked flip side, though, is the positive influence of peer pressure. Teens who befriend academic achievers, for example, will often work to get their own grades soaring.

Adult friendships subtly steer our beliefs, our values, and even our physical and emotional health. Although resolutions to enact new diet and exercise plans and vows to change our character are all too easy to break, if we befriend people whose philosophies and habits we admire, we naturally start adopting aspects of their personalities and lifestyles through a positive desire to be with and to be like our friends. The health-friendship connection is particularly compelling: One study of nurses with breast cancer found that women without close friends faced mortality rates that were four times as high as those nurses with at least ten friends.

The book will explore the "dark" side of friendship, too, to help you understand some of the uglier feelings that come along with amiable affection. Since friends have a hold over us, their power can damage and destroy just as it can heal and help. I'll also tease out the conflicting findings about online friendship to clarify how the latest modes of electronic socializing alter our flesh-and-blood bonds.

Friendship has always been, and will always be, a cherished aspect of human life. But now, just as friendship is rising amid the rescrambling of social structures, we're finally getting a handle on the complex ways that this relationship affects us. Learning how we can get the most out of our friendships is an important endeavor for anyone concerned about well-being, and unraveling the thick narrative strands contained within just one friendship is a fascinating exercise in its own right. The closest of friendships contain the mysterious spark of attraction and connection as well as drama, tension, envy, sacrifice, and love. For some, it's the highest form of love there is.

CHAPTER 1

Deep Roots and a Long Reach: What Is Friendship?

IT WAS TIME FOR HOLLY KILE to leave her husband. He was controlling and emotionally volatile; he wasn't working, and he wasn't helping her run the household, either. "It had gotten scarier and scarier every day," Holly says. Her son was just a year old, and she didn't have enough money for her own apartment. But she did have Erin. "We got into Erin's car," Holly says of that nerve-racking night ten years ago, "and she said, 'We're going to pack your stuff, and you're going to stay with me.' Erin had three kids of her own, yet she never gave it a second thought."

As she settled her baby into the spare room, Holly worried about imposing and being a good houseguest while she was reeling from the breakup of her marriage. "But Erin made me feel like I was part of her family. It didn't strain our relationship at all. I was expecting this awkward transition. That I was totally welcome instead was a profound experience for me."

The two had known each other only since Erin had begun working at Holly's office a few months earlier. "From the moment we met, we were best friends," Holly says. "It was like love at first sight—but not in a romantic way. We instantly knew that we connected on a soul level. That had never happened to me before, and hasn't happened since."

Erin was feminine and fashionable while Holly was a no-nonsense tomboy. They liked different kinds of music and movies, but they somehow recognized in each other deeper compatibilities—the sort of implicit understanding that lets them feel close today even though they live in separate cities and chat at length only a few times per year. "There's an air of love and never any stifling expectations when we talk," Holly says. "It's like having a sister, but one I never fight with."

Sometimes the greatest compliment we bestow on a friend is comparing him or her to family. In such cases "family" connotes total acceptance and the highest sense of commitment and loyalty. But just as often, the way we honor the special nature of a dear friend is to point out how he or she surpasses relatives: "a sister I don't fight with," "the brother I always wanted," "my family of choice."

Friendships are the least institutionalized and most voluntary social relationship we have. Our friends can cycle in and out of our hearts and calendars; they can be our "everything" or just a refreshing anomaly, a small pop of color in a busy social landscape. Amorphous in nature, friendship fills in the cracks left open by our personalities, or backgrounds, or temporary circumstances. Friends adapt to our needs and styles, and we to theirs. Perhaps we'll never arrive at a precise definition, but descriptions of true friends can bring a jolt of recognition. Holly, who was taken in without a word of judgment or hint of hesitation during the darkest moment of her life, says a friend is someone who "knows that you need them, but also knows *how* you need them. Someone who knows exactly how to comfort you."

Pinning Down a Cloud: Classic Definitions

Around 44 BC, Cicero wrote his widely referenced piece on friendship, "De Amicitia." In it, he quotes the Roman poet Ennius:

How can life be worth living, if devoid
Of the calm trust reposed by friend in friend?
What sweeter joy than in the kindred soul,
Whose converse differs not from self-communion?

Compare those touching words with another poem, addressed directly to a friend:

You paint my life brighter
And make my heart lighter

That one is from a Hallmark card. The quality of expression may vary, but some sentiments clearly do not change over time. While Cicero sets a rather high bar of entry into the pal club—"Friendship can exist only between good men"—he upholds a model of the relationship many surely subscribe to today. "True friendship is rare," he writes. "A definition of friendship: complete sympathy in all matters of importance, plus goodwill and affection." A friend, he summarizes, "provides hope for the future," and he claims, "No life is worth living without the mutual love of friends."

The French essayist Michel Eyquem de Montaigne penned

another go-to work on the subject, "Of Friendship," in 1580. Though his particular portrayal of his bond with Étienne de La Boétie is unusually intense, his general insights feel fresh and spot-on, pointing to a deeply ingrained need for—and appreciation of—friendship among humans.

Contrasting two friends to a father and son, Montaigne observes that while friendship is nourished by communication, secret thoughts of fathers aren't meant to be shared with children, and children, for their part, cannot be expected to give proper "admonitions and reproofs, which are one of the first duties of friendship." Brothers, he notes, are close not out of voluntary freedom but out of law and natural obligation. He has a point. Siblings do refer to each other as friends or even best friends, for instance. Yet the very fact that this usage conjures up a picture of an especially close, caring, and fun-loving pair implies that this friendship experience ranks above that of the typical sibling pair. If I say my sister is my friend, she's a sister "plus."

As for friendship among moms and dads and kids—most parenting experts agree with Montaigne that trying to be a buddy rather than an affectionate yet authoritative figure is a seriously misguided tactic. Biological anthropologist Gwen Dewar, Ph.D., summarizes the research on parenting styles and concludes that kids can become overburdened and distressed by a parent who confides in them grown-up worries about finances and interpersonal problems. Some researchers believe that parents who were concerned more with being "liked" as a friend than with being respected as a leader caused the uptick in feelings of entitlement and narcissistic traits among today's young people, compared to the youth of 1979.

Perhaps the sharpest distinction Montaigne draws is that

between love for a woman and love for a friend. (Women themselves, largely uneducated in his milieu, weren't fit for friendship in the writer's view. If only he could see the *Sex and the City* franchise profits now.) Romantic love is "more active and fiercer" than amicable love, but it's also more fickle, he notes. In a nod toward the timeless game of playing hard to get, Montaigne recognizes another crucial difference between the two types of relationships: The more friends want and enjoy our company, the more we tend to enjoy theirs, whereas lovers sometimes become more desirable the more they pull away from us.

The relatively placid lakes of friendship are indeed a welcome respite for many from the choppy seas of dating and mating. Still, these bodies of water intermingle more often than Montaigne imagined. Twenty-first-century partners could be referring to a few possible scenarios when one of them claims, for example, that "my husband is my best friend." Context is needed to parse that boundary blurrer. If the preceding statement was something cheery like "I'm so close to him. We're silly together, even when we're doing mundane things like chores," then it's a safe bet that the man in question is a husband "plus" a laid-back pal. On the other hand, if the label is affixed after a wistful remark about the natural waning of passion over time, you can assume it's code for a husband "minus" romance and/or sex.

Survey Says: What's It to You?

Since Montaigne, we haven't made too much progress in nailing down exactly what makes someone a friend. But the fact that this relationship is not circumscribed accounts for its strength.

Through extensive personal interviews with subjects, Brian de Vries, Ph.D., a professor of gerontology at San Francisco State University, has attempted to grapple with the maddening variety of friendships by honoring the particular meaning that develops between each pair of pals. By doing so, he has mapped out the most common ways people talk about and think about their friends.

A key study analyzed older adults (ages fifty-five to eighty-seven) in both Canada and the United States. The goal? To establish several main categories of friendship. The responses of the interview subjects were classified and counted. One major category settled on by the research team is behavioral processes, which cover things friends do with each other, such as play a sport or have a deep conversation. Another is cognitive processes. These describe how a friend might appreciate or respect you, share your interests or values, and demonstrate trust, loyalty, and commitment toward you. A sense of humor in a friend would also be classified as a cognitive process. The third main category is affective processes, which refer more to emotions than behaviors or ways of thinking. Feeling comfortable with a friend—so that you don't need to explain yourself—is a quality this category captures. The fourth entails demographic characteristics, such as a friend's marital status. You might bond with a friend if you both are recently widowed, for example. And the final category is proxy indicators of friendship (length of acquaintance, frequency of contact, and duration of contacts), for example, "She's my friend because we've known each other for a long time and chat every week."

The results? Friendship, as suspected, is rich and multilayered. The study participants identified seventeen key criteria of friendship but focused more on behaviors than anything else: A friend is as a friend does, apparently. Also frequently cited were cognitive processes—sizing up a friend as loyal or as having shared values and interests. Less frequently mentioned were affective processes, structural characteristics, and proxy measures of friendship.

You might conclude from the study that each person has her own notion or standard of friendship: To Jane, a friend is someone she can trust and to whom she can reveal secrets, while to Joan, a friend is someone who will help her clean out the garage and pick her up from the airport. Instead, it shows that while some factors are more common to friendships than others, there are lots of ways to qualify as a friend.

Because of our own particular talents or interests, we tend to offer a personal "brand" of friendship to people. I'm not handy around the house (or outside of it), so I'm never going to be someone who fixes things for loved ones. I enjoy talking and listening to others, though, so I offer my friends the opportunity to confess fears and sort through their problems, something (I hope!) they appreciate about me—even if they also happen to value plumbing or electrical skills in one of their other pals. De Vries's study tells us that behaviors and shared interests are key ingredients. But the beauty of friendship is that we usually have more than one, and can collect a variety of friends who cover a range of needs.

If I ask you, "What does friendship mean to you?" you might say loyalty or compatibility, in the abstract. However, if I ask you why eight different people are your friends, I'll bet you would describe their individual qualities, the circumstances in which you met, and the traits they tend to bring out in you—this one invites you to fun parties and that one challenges you to be a better person. In other words, asking people to define friendship in the first place is a bit like asking people to define flowers. Friends have baseline characteristics just as flowers are basically the blossoms of a plant, but beyond that they are unique and thrive under very different conditions.

Now, contrast a spouse to that colorful bouquet of friends you're free to arrange and rearrange. A husband is, by definition, the person to whom a woman (or man, in some states) is married. The fact that the relationship is institutionalized and limited to one person leaves no room for confusion or interpretation. Someone could say, "I married my wife because she's smart, pretty, and nice," but the reason she is his wife is that he is married to her.

Many hem and haw over whether they should get married, and to whom. (He's got A and B quality, but what about C?) Meanwhile, if you don't click with your siblings or your own parents, that's a shame, since they are the only ones you've got. But there's no mandate or cultural pressure to settle on the ideal friend for you, and thus there's no need to settle on what a friend means to you.

If the elusive nature of friends bothers us at times, perhaps it's because we're anxiously questioning who our "real" friends are, especially in this age of online contacts (which we'll explore in Chapter 7). If someone wants something from us, can she truly become a friend? What do we call a friend from the past who wants to indulge in nostalgia without recalibrating to the present? Or someone we see every day at the local deli, whom we're really fond of, but certainly not close to? I say it doesn't really matter. Your amicable relationships form a matrix: You're close to some and not close to others; you might like some you're not close to a bit better than others you are close to; you used to be close to people you're not close to now, and you're now close to people you used not to be close to. Each friend has a (nonpermanent) place. When pondering friendfluence, it's useful to decide who your real friends are by using your intuition and self-knowledge.

Friends Around the World

In his book *Friendship: Development, Ecology, and Evolution of a Relationship*, anthropologist Daniel Hruschka, Ph.D., of Arizona State University, takes a fascinating world tour of buddies, all the while illustrating how friendship shifts slightly with context. "Depending on the culture, friends share food when it is scarce, provide backup during aggressive disputes, lend a hand in planting and harvesting, and open avenues of exchange across otherwise indifferent or hostile social groups," he notes. In certain societies, people become friends with those their parents or communities have selected for them. These bonds are sometimes made official through public rituals that make it harder to mistreat a friend going forward. In central Africa, for instance, the Azande exchanged blood to seal friendship bonds. They apparently believed that "a friend's blood stays in one's stomach and will become poisonous if the friend is betrayed."

Americans are evidently less likely than people of other nationalities to lie for their friends. Specifically, Hruschka reports, fewer than one in ten managers in the United States said they would lie under oath to protect a buddy, whereas in Japan and France about three out of ten would. A full seven out of ten Venezuelan managers would fib for a friend. Another interesting (intranational) difference emerges between white Americans and East Asian Americans: Receiving explicit verbal support from friends is a stress reducer for whites but makes East Asians *more* stressed and agitated.

Still, Hruschka concludes that friendship differences across cultures are small relative to the similarities. He formulated the following definition of friendship: "A friendship-like relationship is a social relationship in which partners provide support according to their abilities in times of need, and in which this behavior is motivated in part by positive affect between partners. A common way of signaling this positive affect is to give gifts on a regular basis." Presents, it seems, are the great global equalizers.

Built for Buddies

The females recently studied in a nomadic group tend to hang out in one another's temporary homes. Over the years, they prefer certain companions to others and reconnect with those pals after separations. They gossip and exchange information about the group's planned moves. Some even serve as midwives, fanning and comforting one another when they give birth.

Sounds like an unsurprising finding, except that hanging out in this group consists of literally hanging upside down, side by side. The females in question were bats, and bats, it turns out, have friends, too. Gerald Kerth, Ph.D., the head investigator, noted that while other animals with "friend" types of relationships, including elephants and dolphins, have large brains, his study shows that even creatures with "peanut-sized brains can also have long-term relationships," since they can still keep track of their relationships over time.

A recent study of macaque primates points to how friendship shapes cognitive and social skills. These particular macaques follow a friend's gaze more quickly than the gaze of a relative. Gaze following is an important sign of development because it is a method that animals (and people) use to get information about their surroundings and to learn to communicate.

When animal and human behaviors overlap, as of course many do, it's often a clue to how deep-seated those behaviors are within people. It's clear that people have long had the urge to make friends. From an evolutionary perspective, that tendency is, on the surface, a puzzle: If perpetuating our own genes is life's primary objective, why do we develop such intense nonsexual interest in people who are not related to us? Psychologist Michael Tomasello, Ph.D., has done groundbreaking studies with both chimps and toddlers supporting his theory that the ability to cooperate—a likely precursor to friendship formation—is what sets humans apart from the apes.

"Humans putting their heads together in shared cooperative activities are thus the originators of human culture," Tomasello writes in his book *Why We Cooperate*. "How and why all of this arose in human evolution is unknown, but one speculation is that in the context of foraging for food (both hunting and gathering) humans were forced to become cooperators in a way that other primates were not."

You might think families would have protected one another from enemies, bad weather, starvation, and so forth in the evolutionary environment, but Daniel Hruschka writes that kin just weren't enough. First of all, relatives probably have access to the same resources that you have access to, rendering their contributions somewhat redundant. Second, for big projects, one family can't always provide enough manpower to carry out the job. Finally, people are sometimes thrown into situations in which they have no relatives nearby.

In our distant ancestors' environments, evolutionary psychologists argue, women tended to leave their families and join their husband's tribe, making it important for them to be able to form ties with nonrelatives. Otherwise, they wouldn't be able to secure the help they needed to survive and raise their children successfully. This may be an explanation for what some see as women's greater emotional investment in close friendships. Men, for their part, would have also needed friendship skills in order to form alliances and obtain power and protection within the tribal hierarchy.

Most of us don't count on our friends for survival these days, but because we're made to put utmost mental and emotional energy into building bonds and staking our place in the warm safety of a group, sometimes it sure feels as if we'd die without them.

Short Friendship, Long Impact

The "proxy indicators" that some (especially men, as de Vries found) use to define friendship—how long they have known the person and frequency of contact—don't necessarily correlate with friendfluence. Steve Weitzenkorn, now sixty years old, met Jerry his freshman year at Ohio University, where Jerry was a savvy sophomore. "I was a shy, insecure kid, and Jerry was assertive and knew how to connect with professors and graduate assistants," Steve recalls. Soon Jerry roped Steve into visiting the psychology lab where he was helping with some research projects, mostly in the area of achievement motivation. The fact that Jerry got himself involved in these voluntary endeavors was a revelation to Steve. "This was a whole new skill-set for me because I wasn't the type to go after opportunities. He helped me climb out of my shell, learn by doing, become more confident, and develop a far deeper knowledge of my discipline."

For two years on campus, the pair was inseparable. But soon other friends started to capture Steve's attention. "The downside of Jerry is that he had a strong personality. He was very intellectual. He loved to debate. And I found that the other people I wanted to hang out with on campus were turned off by his argumentative side." He and Jerry drifted apart and then went to graduate school—Jerry in Georgia and Steve in Nebraska, where he earned a Ph.D. in organizational psychology.

The two got together once in a while, but the intervals between meetings grew longer. "We got wrapped up in our own lives." Still, Steve continued to live by Jerry's code. "I applied the lessons I learned from him in being proactive and taking healthy risks . . . in various jobs, entrepreneurial initiatives, and community activities. They are also lessons I have passed on to my son, who applied them in his own college career by working at the campus radio station and discovering what he's passionate about outside of the classroom."

Steve did a Google search on Jerry three years ago in hopes of getting back in touch. The first result on the screen was Jerry's obituary. "I was shocked. I regret that I never really told him in explicit terms how grateful I was for his help and how it influenced me," Steve says. "I have one of those old-fashioned circular Rolodexes, and many years ago, I put his card in there. Occasionally it pops up as I'm flipping through, and it reminds me that he was my friend and that he made a difference in my life."

Bonding Buddies and Gabbing Girlfriends?

That it was based on shared activities—working at the lab—and lacked declarations of how they felt makes Steve and Jerry's friendship typically male. But Geoffrey Greif, D.S.W., professor at the University of Maryland School of Social Work and author of *Buddy System: Understanding Male Friendships*, warns that though some gender differences in friendship can be tracked, they don't apply to all men and all women.

Greif interviewed about five hundred people about their social lives past and present and identified key style differences of same-sex friendships. "Men tend to have shoulder-to-shoulder friendships and women face-to-face friendships," he writes in one summary of the findings.

Eighty percent of the men Greif interviewed "said they participate in sports with their friends; no women gave that answer though a few said they exercise with friends. Shopping is a more common activity for women—only one man out of 386 said he shopped with his friends."

Women are more likely to reach out to friends in order to keep up the contact, Greif found, while men are less likely to feel upset or take action if they've fallen out of touch with a pal. Women tend to show support by listening, and men show support by offering advice. Perhaps because they are afraid of seeming gay, Greif reasons, or because of the way they have been socialized, or even because of biological factors, men are less emotionally and physically expressive with their friends than women are.

In spite of these surface differences, Daniel Hruschka concludes that the spate of recent claims about the way that men's and women's friendships are different is contradicted by the hundreds of studies showing that the two sexes "are actually quite similar in how they engage with friends." Men and women both say that they rely on close buddies for companionship even more than they rely on brothers and sisters or parents. Both sexes hold similar expectations for close friends and have both "special-purpose" friends, with whom they engage in one activity, for instance, and "multifaceted" friends they can take almost anywhere. Apologies to adherents of the "Mars" vs. "Venus" philosophy; it turns out we are all earthlings.

Pop culture portrayals of male friendships, despite their veneer of hipness, adhere to standard gender roles. Witness the "bromance," the idealized but also infantilized relationship between straight buddies explored in Judd Apatow comedies, where video game consoles nurture closeness and the advice, while free-flowing and generally bad, is a sweet assurance that these overgrown boys have got one another's backs.

Men who conduct more "face-to-face" relationships, say, by going to a quiet restaurant together to talk and catch up, received a new label for their gender-atypical actions in 2005, when Jennifer 8. Lee, a *New York Times* reporter, coined the term "man date" for "two heterosexual men socializing without the crutch of business or sports." The men interviewed for the piece alternately defend the value of their beer and wings–free bonding time and confess their preoccupation with appearing gay.

Maybe I have namby-pamby male friends, but I know plenty of guys who analyze life and each other more seriously than Apatow's characters do and who would never feel uncomfortable being seen having dinner together. In fact, my (perhaps romanticized) conception of the man date is more macho than effeminate, harking back to the time when women couldn't or weren't allowed to engage in deep conversations about politics or culture and when men retired to drawing rooms, brandies in hand, to do so.

Harry, Sally, Whatever

In 2009 the University of Chicago instituted a new housing policy allowing men and women to room together in the dorms, "something that was forbidden throughout the 117-year history of the Hyde Park school," the *Chicago Sun-Times* reported. One female student who took advantage of the new option by recruiting a close male friend to bunk with her said, "I tend to get along better with guys." Katie Callow-Wright, the university's director of undergraduate student housing, said that there had been "no complaints, no issues, no concerns" over the cross-gender living arrangements.

This is a nice data point in support of the notion that in spite of the Harry and Sally Problem—code for "men and women can't be friends"—cross-gender friendship is certainly possible and increasingly accepted. These can be particularly helpful and valuable ties, despite the tensions and difficulties they might also contain.

Lucy Taylor, a British journalist, starts out a first-person account of why women should have male friends by detailing a recent camping trip she took with her best friend, Andy. "He taught me how to fish and build a fire. We spent the evening watching the flames, eating our (OK, his) catch, drinking wine and talking about life, a mutual friend from our schooldays who'd recently died, our work and a bit about our relationships, then we went to bed, sharing a tent, though not a sleeping bag. Yes, he's a man and I'm a woman. We're both heterosexual. We're very good friends. And we're not doing it—never have and, unless the proverbial pigs start flying, never will."

Taylor, perhaps for dramatic effect, describes an extreme degree of togetherness that would understandably make romantic partners jealous. While she insists her boyfriend and Andy's girlfriend think the camping trip in particular and the friendship in general are kosher, some of her other friends "find this weird, unbelievable, and highly suspicious." Taylor concedes that many marriages have been broken up by opposite-sex friends and that men are more open with female friends than with male friends, paving the way for an emotional connection to become a sexual one. But, she argues convincingly, that shouldn't make these enjoyable friendships forbidden. In contrast to her female friendships that sometimes devolve into petty fights and "bitchiness," her relationship with Andy is refreshingly envy-free. Taylor is grateful for Andy's male perspective and thinks others should get the benefits of cross-gender friendship without judgment.

Around the same time that Taylor's manifesto was published, a fellow British journalist, Sarfraz Manzoor, penned his own essay on why he enjoys female friends. "I was not allowed to have girlfriends or girl friends when I was young. My parents were traditional working-class Pakistani Muslims who strongly disapproved of boys and girls socializing together," he explains. "I was painfully shy around girls throughout my teens and 20s, and it was not until I left home that I began making female friends. This was partly because I was rubbish at romance; I found it frustratingly difficult to get girlfriends but very easy to have girl friends." Manzoor's female friends helped him navigate his crushes and eventually his real romantic relationships. (One helpfully warned him not to reveal his hypochondria by declaring a mild headache to be an aneurysm while on a date.) "I was dating one woman for more than a year before I mentioned it to my closest male friend. Maybe it is male pride that keeps me from opening up to my guy friends—I don't want to reveal my vulnerabilities and insecurities," he writes.

Just as they give him opportunities to open up and ask for

help, Manzoor (now married himself) levels with his female pals and gives them the honest scoop on their situations: "I will be asked to decode the latest phone conversation/text/email. She: 'He hasn't called when he said he would . . .' Me: 'He may be really busy or just a bit shy.' She: 'But he isn't that busy.' Me: 'In that case maybe he just isn't into you.' "

"The most important thing I have learned from my female friends," Manzoor concludes, "is that men and women may be very different, but each has the ability to bring out the best in the other."

Young adults seem especially comfortable with cross-gender friendships. One survey found that eighteen- to twenty-four-year-olds are nearly four times as likely as people over fifty-five to have a best friend of the opposite sex. In contrast, in a study that explored the role of friends in the lives of married couples (an older cohort), not a single person among the 654 participants declared that he or she had someone of the opposite sex as a best friend.

Perhaps the "best friend" distinction was too exclusive. When the authors of *Friends Forever*, Suzanne Degges-White, Ph.D., and Christine Borzumato-Gainey, Ph.D., conducted their own qualitative survey of married or partnered women, they were surprised to discover that many of the women claimed men as close friends. Backing up Taylor's personal epiphanies, these subjects tended to note "the lack of drama, the lack of competition in the friendship, and an ease in sharing and companionship as the advantages over friendships with other women."

Cross-gender friendships are hard to define and hard to defend to others who regard them suspiciously. Also, sexual attraction does frequently come up as a challenging factor to contend with, even if the friends are very capable of keeping any flutters of lust in check. (Men report more attraction to female friends than vice versa.) The rewards of these friendships seem to outweigh the obstacles, however. One study in fact found that both men and women rated these friendships as higher in quality than same-sex ones.

Better than Sex!

If there is sexual tension between cross-gender friends, how often do they end up sleeping together? And does that make them different from friends who keep their clothes on? In 2000, Penn State researchers found that "approximately half the heterosexual college student population has engaged in sexual activity in an otherwise platonic cross-sex friendship." With friends like these, who needs lovers? But this eye-popping percentage is likely specific to college students, among whom "friends with benefits" is apparently a very accepted category. Interestingly, though, more than half of those who did have sex with a friend did not then fall into romantic relationships. The implication is that the friendship was more important to them than the sex.

In 2010, Justin Lehmiller, Ph.D., and colleagues at Colorado State University decided to reach beyond the campus and recruited an Internet sample of 411 people (average age twenty-six) who were having casual sex with at least one friend. Lehmiller was interested in gender differences: Would the women be more keen to transform a casual arrangement into a formal romantic one? Yup: "Men were more likely to hope that the relationship stays the same over time, whereas women expressed more desire for change into either a full-fledged romance or a basic friendship." But what surprised the research team was the finding that both men and women were "more committed to the friendship than to the sexual aspect of the relationship." Corroborating the Penn State study, the researchers concluded that "regardless of partners' sex, friendship comes before 'benefits' in friends-with-benefits relationships."

Again, we see how friendship, in all its flexibility, morphs around circumstances. The essential "friendness" doesn't evaporate when two buddies sleep together. But if they start to fall in love (and the mere act of having sex can trigger romantic feelings, thanks to the bonding hormone oxytocin, which is released during orgasm) and become a couple, are they still friends? Or are they romantic partners who began as friends? Friendship and romantic love are not mutually exclusive. For the purposes of examining friendfluence, though, we can agree that even if friendship is at the core of a romance, once people are formally partnered up, they are no longer "just friends."

Friendly Margins: Gays and Lesbians

When Peter Nardi, Ph.D., professor emeritus of sociology at Pitzer College and author of *Gay Men's Friendships: Invincible Communities*, began researching the topic of his book in the '90s, he learned that not many others had looked into it. "But what I had always heard from gay friends was 'friendship is family,'" he says. "It wasn't an option for some; it was a requirement, since many had been ostracized by their own families."

Nardi discovered that indeed, for many gays and lesbians "friends as family" was not a metaphor but a reality. Friends spent holidays together and formed tight communities. Through surveys, Nardi identified specific functions of friendship in these groups. First of all, friends aid identity development: "Unlike heterosexual men in general, gay men are more likely to talk to friends about issues of identity." Chatting about such deep concerns can in turn lead to greater intimacy among these friends. Friendship also serves as the basis of political involvement for gays. "Whether it's Prop Eight or any social change movement, it grows through friendship networks," he says. Thanks to the Internet, which enables young gays coming out to read about and get in touch with others in their situation, identity issues are probably easier to weather these days, Nardi speculates. "Still, there is plenty of gay bashing and suicides, so the need for friends to help each other remains crucial."

More recent work from de Vries indicates that gay men and lesbians define friends on the basis of their actions more than heterosexuals do. They were also more likely than heterosexuals to cite similar beliefs and strength of feelings for one another as well as proxy measures of friendship. De Vries interprets this as a marker of the depth of consideration with which they assess their friends. Since they might have supported each other in times of being shunned by their hometown friends or being discriminated against at work, they figure prominently in each other's lives.

Shane Allen, twenty-four, and Felipe Baeza, twenty-three, met four years ago at Hecho en Dumbo, a restaurant where they both work as they pursue classic New York City dreams. Shane is an actor, and Felipe is a visual artist. Both possess sweet, upbeat personalities, but their preferred mode when speaking of each other is sarcastic teasing. "I thought Felipe was a not-that-bad-looking Mexican guy," Shane says of the time he first spotted him on the job. "We started going out after work," Felipe adds. "Shane would get messed up and I would take care of him."

A period of shared poverty fortified the bond. The restaurant closed down for months while the owners moved it to a better location. "I had saved money for three months," says Felipe, "but the renovations took six."

"I didn't have anything saved," says Shane.

"It was winter, and we were freezing, looking for jobs and getting turned down all over the place," says Felipe. "We were really hungry."

"And there was that place by your house, remember?" says Shane. "It was El Salvadoran."

"Yeah, we ate there every day. They sold *sopes* for seventy-five cents," says Felipe.

"We would crouch over the counter and share one with our hands. It was pathetic," Shane says, laughing.

The hard times are over, but their friendship remains strong. When they're not hanging out at work, they talk on the phone, and they like to go to museums on Mondays. "We help each other with emotional stability," says Felipe.

"Felipe is a mediator between me and my boyfriend," says Shane. "I'll start arguing with him, and Felipe will calm things down. Everything is a problem with me. And everything's not a problem for Felipe. He's chill, and I bring the drama."

"Shane's crazy, but in a good way," says Felipe. "Maybe that's why I like hanging out with him. I'm not that crazy, but *he*'s my crazy."

Shane and Felipe are part of a younger generation that is generally more tolerant than older generations. Because they have not personally experienced acts of discrimination, or rejection from their hometown family and relatives (with whom both are close), Felipe and Shane don't "need" each other to replace their family members or to protect them from any hostile forces, but they have relied on each other for the old-fashioned support everyone needs to make their way in the world, and to manage coming out and same-sex romance.

"When Felipe and I became friends," says Shane, "I hadn't dated many guys yet. It's nice to have someone who can relate to me as I go through that. And with regards to being gay, I'm less stressed out about it because of Felipe. I feel normal. And I feel like, whatever happens, everything will be fine."

DE VRIES BELIEVES THAT FRIENDS are very important to social life. Yet, as he writes in one of his papers, "The benefits of friendship are unfortunately constrained by a culture and social system in which friendship is frequently discounted or doubted—or worse. . . . In contrast [to families], the ambiguous social and legal recognition of friendship dramatically limits friendship's potential." It might not affect your personal constellation of friends much, but at the big-picture level, how it is defined has real consequences on how friendship can flourish and contribute to the well-being of citizens and the smooth functioning of society.

Consider the health care system, which de Vries says doesn't recognize the role of pals. Research on friends who are caregivers points to the contortions they often have to go through to convince staff members and the patients' relatives who are not caring for them that their intentions are pure. That's because there is a presumption that a friend would be more likely to take advantage of a patient than a family member would, he says.

A tendency to minimize their importance is also apparent in the way we treat someone whose friend has died. "I call them disenfranchised grievers," says de Vries. "No one ever sends friends condolences or flowers. No boss would give you a week off if your friend died. No airline would give you a compassion fare to travel to a friend's funeral. Only the family is entitled to grieve."

Often, mourning a friend is just as painful as mourning a relative, yet the lack of formal support for these grievers can make it harder for them to cope. Just as we can learn about friendship's possibilities by looking at people who make the most of their friendships, we can learn about friendship's meaning through grief. "As a culture, we're less likely to say 'I love you,' to a friend," says de Vries. "When friends die, we often realize how much they mean to us, and how little we communicated that to them."

ANY SUBJECT LONG EMBRACED BY philosophers and poets is bound to be an intriguing and emotionally provocative one; friendship easily merits those descriptions. Social scientists' systematic approach might drain friendship of some of its elusive magic but it does expose the most paradoxical quality of friendship: It's the most stable—in that one is likely to have a friend at every point in life no matter what other relationships one has or doesn't have—and it's also the most flexible.

CHAPTER 2

Finding and Making Friends

IN JUNE OF 1975, A GIRL WAS BORN in the dense capital city of Lima, Peru, to a nineteen-year-old woman living in a large house full of her relatives in a neighborhood of imposing mansions. A few months later, another girl was born in the small town of Rutland, Vermont, to a thirty-three-year-old woman living in a two-story colonial with her immediate family, on a quiet street with a view of purple mountains in the backyard. Sure, we're the same age, but how could anyone have predicted that the two babies—Sofia and I—would become best friends?

I wanted to befriend Sofia at first sight: I was in the cafeteria of my college dorm when I saw a tall, ballet dancer–thin classmate in a gypsy-style outfit, with brooding black eyes to match. She seemed independent and confident in her quirky style, yet distractedly dreamy, as though she were in another, far more fascinating world. A few days later, she knocked on my door in search of my roommate, with whom she shared a class, and unwittingly launched our friendship.

Sofia and I have seen each other through the process of

growing up. We helped forge each other's interests and outlooks and, painful though it was, exposed each other's defenses and rough edges, in loving attempts to smooth them out. That quiet confidence of hers came from a deep acceptance of her offbeat intelligence, which in turn helped me become more secure in the ways *my* mind worked.

Because she understands me so well yet has also shown me worlds completely different from my own, our meeting sometimes seems fated. We often want to believe that a higher power put our friends in our paths, if only because we cherish them and shudder to think of the counterfactuals: "If I hadn't met so-and-so, I never would have done X or Y!" Social scientists obviously can't prove or disprove divine intervention, but they are getting closer to uncovering how we befriend each other and why we become attached to some acquaintances but not to others.

We might think we "choose" our friends and, as we established in Chapter 1, that *is* what distinguishes them from families of origin. But "choose" is really shorthand for a mysterious process that includes some factors beyond our control, in every sense, and other factors that are at least beyond our conscious awareness. Why was I drawn to Sofia in a visceral way that resembled "falling" for someone romantically? I can't say for sure, but many people describe initial encounters with someone who would become a close friend as powerful emotional events akin to love at first sight. Others, though, feel not much at all but then later attach meaning to the nonevent of a first conversation with a pal—only because the relationship later developed into something important.

Since friends are so influential, it's important to know how we tend to find them.

Friend Formation 101

Half a century ago, researchers came up with the "proximity theory" of friendship—that we befriend people who live geographically close to us or who frequently cross our path because they go to our school, grocery store, office, or favorite diner. Proximity, first and foremost, grants easy opportunities to meet. But also, familiarity breeds positivity. Called the "mere-exposure effect," it's a phenomenon that is widely documented: Just seeing someone over and over can make you like him or her more. It's probably because familiarity feels good to brains that would rather process stimuli using worn-in neural pathways than by forging new ones. The mere-exposure effect lends nearness a doubly powerful advantage.

You may have had the experience of meeting someone who reminds you of someone you already like; positive associations can prime you to want to be around that person, even though you'll probably find out that she is in fact different from the one whose qualities you've already projected onto her.

One friend-making fun fact recently floated up from the research files: You'll give off a better first impression (and thus possibly make friends more easily) if your name is easy to pronounce. It will be interesting to see if this finding holds up, given the trend toward unique and multiethnic baby names.

Shared activities are fonts of friendship. Not only are you near and familiar to those in your yoga class, but you also have something in common: a love for, or at least a begrudging commitment to, stretching and breathing. Much has been written about the decline of structured recreational activities in the United States (bowling leagues being the iconic symbols of a bygone, socially organized utopia) that enable such friendships. This is caused, some argue, in part by suburban sprawl, which puts people farther away from the center of their communities, and to a certain extent by the colossal amount of time people spend watching TV in the isolation of their own homes. However, others have more recently argued that the Internet has caused a flowering of activity-linked groups that operate online and in 3-D.

Major life events and changes bring on a need for social support and at the same time usher us into settings that spur new ties. Take the new mom, for instance: She's feeling simultaneously terrified and euphoric about the experience of caring for a helpless little blob/miracle. Her childless friends are well-meaning but don't grasp the complexity and contradictions of her new responsibility and lifestyle. The lady she met in her birthing class, however, *does* relate, and so their friendship develops at an accelerated rate. Ideally, the new mom still keeps her old chums, though; after all, they still see who she is fundamentally and can help her hold on to her precaretaker identity.

We find a few characteristics generally attractive in potential friends, notes Daniel Hruschka in his anthropological and psychological survey of friendship. "These include a reputation for helping, high status, and interpersonal similarity in social class, ethnicity, and personal attitudes." Also, just knowing someone likes you makes you like him or her more.

Back in 1937, Dale Carnegie published his hugely popular book *How to Win Friends and Influence People*. Carnegie urged people to copy the behaviors and traits we naturally find desirable in others, such as smiling a lot, encouraging conversation partners to talk about themselves, and using their names frequently. Research has since shown that these tactics will indeed warm others up to you. Even invoking the word "friend" can prime others to feel friendly toward you. When running for president in 2008, Senator John McCain used "my friends" every four minutes, on average, during a debate that went on for an hour and a half. (Now, the fact that he didn't win might discount the suggestion, but perhaps he made many pals along the way.)

We gravitate toward those with strong social skills simply because talking to them is less work than it is with closed-off or awkward types. This doesn't mean that just because someone is a talented conversationalist who puts you at ease he will become your best buddy. But like proximity, affability kindles the potential for friendship.

That's why people with Asperger's syndrome, for example, struggle to make friends. Even though they function well in the world generally and are as talented, smart, and interesting as the next guy (if not more so), many report such difficulty engaging in small talk, reading facial expressions, picking up on cues of interest, and falling into the give-and-take rhythm of satisfying conversation that they can't clear the hurdles that precede the true "getting to know you" phase.

Once a conversation is flowing, protofriends must begin the dance of disclosure, revealing tidbits that they wouldn't broadcast to just anyone. Once we do disclose a feeling or experience to someone, we are more attracted to him or her—perhaps in an effort to justify the risk we've just taken by spilling a sad story about our past or voicing a controversial opinion that we hold dear. The act of disclosure itself signals an interest in moving from acquaintance to friend, and then the specific nature of revelations made to us helps us decide if we want to get closer and hear more. As with so much in psychology, striking a middle ground is best when it comes to opening up to someone you are getting to know. Telling your childhood traumas and deepest darkest desires to a fellow cocktail party guest will put you in the "colorful character" or "crazy" category—not the "possible new friend" box.

Then begins the process of building a friendship, a somewhat unconscious one (at least compared to the strategizing employed in the development of many romantic relationships). As Hruschka puts it, budding friends move from "giving tentatively and keeping accounts, the common mode of interaction between acquaintances and strangers, to sharing at high levels of trust and support, without keeping track of past favors." As they do so, they figure out how to communicate more efficiently, read each other, and resolve conflicts. "In studies where people watched their friends in a conversation, they were 50 percent more accurate at inferring their friends' self-reported thoughts than they were at inferring the thoughts of strangers. In short, friends are experts at how their partners think and feel, while strangers are novices."

One way to deepen a friendship is to expand the ways in which you interact with someone—to cultivate what researchers call "multiplexity." The word conjures up an image of going to a movie together at a big theater, but it just means diversifying your activities in order to feel closer and get to know each other better. If you always meet a friend for coffee near your school, for example, invite her to a family dinner (and give her the background on the unique dynamics she'll pick up on).

The longer you are friends with someone, the more likely you'll continue to be friends. Time spent as friends is the best predictor of friendship longevity.

Birds, Feathers, Flock

Oprah and Gayle, America's favorite BFFs, exemplify an important ingredient in the friend formula. Though one of them has reached stratospheric heights, both Winfrey and King are very successful journalists. In fact, they met on the job, at a Baltimore news station, in the late '70s. A *New York Times* profile quotes King as saying: "It's very nice to have someone who really gets you—*really* gets you. Our sensibilities are very much the same. Just the other day, we had a conversation, someone was talking, we both at the same time said, '*Really?*' The same inflection. Our brains are wired very similarly."

Winfrey, in turn, tells the paper: "We are so much alike. That's why we became friends. It's easy for people to think she's trying to be like me, when in fact she isn't. We're just very much like each other. That's what happens with friends: over a period of time, you start to sound alike, and your inflections are alike."

What could be better than being best friends with . . . yourself? But before you assume that these two have come down with fame-induced narcissism, note that we all do this. Attractive young women hit the town with other attractive young women. Skateboarders chill out with fellow skateboarders. Moguls dine with moguls.

"The level of similarity between two potential friends is directly proportional to the chance that a friendship will be launched," write the authors of *Friends Forever*. These similarities include gender, ethnicity, attitudes, beliefs, and values.

Sometimes similarity and disclosure can combine to forge a deliciously affirming friendship. Imagine you're sitting in a mandatory meeting at work, led by a blowhard you can't stand.

During a coffee break, a new coworker whispers to you that she would rather poke herself in the eye than sit through the rest of his presentation. Voilà, a friendship blooms. Jennifer Bosson, Ph.D., a psychologist at the University of South Florida, studies the power of shared negative attitudes and concludes that people connect easily when they both dislike a third person—even if he or she is someone neither of them actually knows, like Kim Kardashian. The coworker in the above example shows herself to be like-minded and also trusts you enough to reveal a potentially risky stance.

Another intriguing new line of research is tracking whether groups of friends might even share genes—an indication that our similarities could run very deep indeed and that the forming of friendships could be, in part, governed by fundamental biological processes and signals. Using data showing subjects' friendship ties and samples of their DNA, James Fowler, Ph.D., and colleagues looked for markers in six different genes and found that pals had a much greater than chance likelihood of sharing a variant of dopamine receptor D2, which interestingly enough is also associated with alcoholism. Fowler, a social network whiz in the political science department of the University of California, San Diego, concedes it could just be that people who like drinking and hanging out in bars will befriend each other before they'll befriend teetotalers. But he continues to look into the matter and even theorizes that our friends' DNA could trigger alterations in our own behaviors and characteristics, a phenomenon documented in the bird world: Hens' feathers change depending on the genetic makeup of the hens in nearby cages.

Yet research does not show that friends are particularly alike in personality, granting scientific credibility to hundreds of romantic comedies wherein the uptight leading lady has a free spirit for a sidekick and the charismatic main man has a buffoonish buddy. Sofia is an introvert, whereas I'm more of an extravert, for instance. That's a big reason we don't see each other as carbon copies. Still, our demographic profile and outlook on life mostly match.

Even if you don't think your friends are just like you, it's doubtful that your gang resembles what we see featured in government brochures or Benetton ads: folks of different ages, classes, and races happily cavorting together. From your up-close vantage point, you can see detailed mosaics made up of the life histories, temperaments, dreams, and quirks of all of your friends. From afar, the world likely sees a uniform flock in motion.

Economists at Dartmouth College used the college dorm—an excellent experimental setting—to watch how friendships actually formed over the course of a school year. They tracked the number of e-mails each student sent to each other student via the internal school messaging system (they'd shown in other experiments that closer friends send more messages to each other than to less close ones). Since students were randomly placed in dorms and with roommates, the factor that interested the researchers the most was race. Would a white student with an African American roommate have a more diverse group of friends by the end of the year?

At a very competitive school like Dartmouth, you might think the fact that all the students are intelligent and serious about academics would render race a less important factor in friend selection. Not so. An African American student and a white student who roomed together did have a greater chance of being in contact with each other via e-mail, but their rooming together did not increase the chances that the white roommate would have additional African American friends. Being in the same dorm made any two students much more likely to be friends than students living in different dorms, but a racial divide still persisted even among those most likely to run into each other and spark a friendship. As one possible explanation, the authors point to neuroscience findings suggesting that white-black interaction is more stressful than within-race interaction. It's a bleak conclusion for anyone who thinks college is a great place to make friends who are different from you: African Americans shared about 44 percent of their e-mail volume with other African Americans, while non–African American students shared about 4 percent of their e-mail volume with African Americans.

Why do we seek out people like ourselves, then? Three common explanations all have some empirical support, and each addresses a different aspect of our internal makeup. The first builds on the familiar-is-good cognitive psychological principle and posits that being around those who do what we do validates our choices and leads to relatively comfortable and conflict-free interactions. The second is an existentialist philosophical view: Finding a soul mate who possesses our attitude toward life and who seems to truly "get" us, as Gayle King put it, wards off the desperate isolation that is a hazard of being human. The third is a page out of the evolutionary biological book: Because our main drive is to propagate, we are drawn to friends who are similar to us genetically. If we stick with our "own kind," someone in our tribe is bound to pass on some of those genes we have in common, even if we ourselves can't manage to reproduce.

The Deeper, Darker Logic of Friendship

At a minimum, we have to be near people in order to befriend them (though these days, "near" could mean a few inches away in our in-box or Skype window). We usually have to be similar to them in some respects, and we often have to possess the necessary social skills to reveal ourselves and first get a sense of whether or not we want to take the leap from stranger to acquaintance and then whether to take the journey to friend or even close or best friend. What exactly moves us along that path toward greater intimacy?

For decades, evolutionary biologists believed friendship to be a function of reciprocal exchange; as long as the back-scratching escalates at a mutually beneficial rate, the friends get chummier. In the late '90s, psychologists John Tooby, Ph.D., and Leda Cosmides, Ph.D., began exploring other reasons we've evolved to be altruistic—why we help others even at our own expense. Tooby and Cosmides noticed that most people got angry when they heard the standard explanation that friendship is a tit-for-tat enterprise. They denied that their own friendships were based solely on favor trading, and they insisted that they greatly enjoyed lending a hand to those they care about. "Indeed," the researchers write in one of their papers, "explicit linkage between favors or insistence by a recipient that she be allowed to immediately 'repay' are generally taken as signs of a lack of friendship." The accepted theory seemed to explain successful business arrangements, not buddies.

In fact, a few experiments have shown that friends care *less* about inequality (in terms of favor granting, etc.) than do strangers or acquaintances. This would seem to make them susceptible to exploitation by friends, so why would they take the risk, especially if the "survival of the fittest" governed the development of friendship and other relationships?

Working within the framework laid out in Chapter 1—the idea that our penchant for making friends has been ultimately in our own interest because comrades have kept us alive—psychologists have proposed that cold machinations lurk behind our warm, fuzzy, and forgiving feelings toward friends. Most interesting is how much more complex these (often unconscious) social calculations are, especially compared to that old tit-for-tat theory.

Consider the friend niche limitation model: You have a small number of close friendship slots, and as someone who is, on a fundamental level, trying to survive in the world and avert disaster, you need to fill those slots with people who are capable of helping you out in unforeseen circumstances. Think of your friends as a balanced stock portfolio and you as the analytical money manager.

On what criteria do you base decisions on whom to let in or out? Tooby and Cosmides suggest that we evolved to include those with special skills or traits, such as someone who is a superior "wayfinder, game locator, toolmaker, or who speaks neighboring dialects." This type of girl would surely have been a good one to have on your team, regardless of how altruistic she was. And it's easy to imagine her modern counterpart. Who wouldn't want a friend who has a GPS-like sense of direction, knows all the best restaurant options, can pass on useful iPhone apps, and speaks Spanish?

You've also evolved to want at least one person in your portfolio who reads your mind fairly well. Those who can guess what you're thinking and feeling are valuable because they are attuned to your needs. Also worthy are those who want the same things you want. Such people will work to "make the world suitable for themselves," and thereby for you, too. And finally, you'll tend to fill a friendship slot with someone who considers you difficult to replace. "This person has a bigger stake in your continued health and well-being than an individual who can acquire the kind of benefits you provide elsewhere." That explains the inevitable popularity of the kid in the neighborhood with the pool table.

Get Your Allies in Order

That we track how others might assess us as they examine their own friend portfolios is the key to the latest idea to grow out of evolutionary psychological principles: the alliance hypothesis. While basic rules of friendship formation explain why someone might become a friend rather than remain a stranger, Peter DeScioli, Ph.D., and Robert Kurzban, Ph.D., professors of psychology at the University of Pennsylvania, point out that for most of their history, humans lived in small groups where strangers didn't generally exist as a concept. We already knew everyone in close proximity, and they were all fairly similar to us. "If human friendship is strategic," they write, "its sophistication will not be found in how strangers become friends, but in how people sort known individuals into best friends, less friends, and enemies."

Whether it's an annoying inconvenience, like a gripe with a negligent landlord, or a spirit-crushing fight, like a years-long custody battle, whenever we get into a dispute, we love to tell other people about our predicament to get them to see things from our perspective and not that of our enemies. Drawing on game theory in international politics, the alliance hypothesis holds that friendship is caused by cognitive systems that function to shore up alliances for when we'll need them. "You want allies before the dispute begins, but you can't anticipate when that will be," DeScioli says.

The main prediction of this hypothesis is that because your friends' ability to support you is compromised if they have other best friends, you will get closer to people who rank you high on *their* list, too. Even our most entrenched disputes are not usually life and death. Yet our brains developed to prepare us for situations in which we might get only one lifeline. The person we expect to throw it to us had better not be obligated to save somebody else first.

DeScioli and colleagues' most compelling evidence is their analysis of the largest data set on friends ever studied. They collected a sample of 11 million MySpace users (back when it was the most popular of the online social networks) and then zoomed in on approximately 3.5 million who named best friends who were also on the site, through a "top friends" ranking feature that all could see.

While proximity had some predictive power—people chose geographically closer people as top friends—the best predictor was the variable they were most interested in: Sixty-nine percent of people selected for a best friend someone who also ranked them number one.

The transparency of the MySpace system notwithstanding, we usually aren't directly told where we fall in our buddies' rankings. An important function of the cognitive systems that drive friendship, then, DeScioli says, is to discreetly monitor and divine our place in others' hearts. This is why you'd be very interested to hear that upon discovering that he is double booked, your friend Chris is going to Bob's party instead of yours. Or why, after getting a call from a friend who announces she is pregnant, you might take the opportunity to put out a little probe: "That's great! Did you tell anyone else yet?"

The do-or-die nature of friend rankings, according to the alliance hypothesis, is why finding out you've slipped down a step is hurtful and can bring on intense feelings of sadness or jealousy. Weddings—ceremonies that often require a bride and groom to publicly rank loved ones via official appointments such as "maid of honor"—frequently expose the ugly side of friendship. Take, for example, two messages posted on TheKnot.com, a popular hub for brides-to-be and those in their orbit:

> My friend got engaged in July. We are best friends and have been since kindergarten. I have been there for her so many times when she was sad, lonely and stressed.
>
> She had an engagement party over Christmas, but it was only for family and the wedding party. I saw pictures that another bridesmaid posted, and it's clear who the bridesmaids are: her sister, and 2 of our good friends. I have no idea why I wasn't asked. I called my friend and said nicely: "I saw your pictures on Facebook. Is it safe to assume I'm not a bridesmaid?" She told me she's sorry if she hurt my feelings, and in no way does this affect our friendship, she just thought it would be an inconvenience/headache for me. I am crushed.

> Have any of you experienced friends/family upset/ disappointed for not having them in your bridal party? I have a friend who is a little upset she isn't going to be a bridesmaid. I can deal with that, I knew some people

> would not be happy, but she's gone so far as to tell me she does not want one of my best friends as a bridesmaid. She says she will be angry if my other friend is a bridesmaid and she's not. How do I deal with this other friend and her jealousy issues? The last thing I want at my wedding is drama.

The jilted pal in the second testimony goes so far as to sabotage the bride's other friend—a typical reaction of those whose rank is threatened, says DeScioli. She could have tried to strengthen her relationship with the bride to win her back as a key ally, or she could have, in soap opera fashion, attempted to knock out her rival by spreading rumors or in other ways damaging the relationship between her friend and the despised bridesmaid. "These reactions look insane from the outside because they are such small matters!" says DeScioli. "But understanding why these events are so important to us can be very enlightening. It's nice to know the logic that drives our emotions. It makes us realize why these problems are so hard, and can maybe even remind us that we don't have to take ancient feelings so seriously in the here and now."

Identity Abettor

Now that we've rolled around in the unconscious mud of rivalries, insecurities, and the manipulative maneuvers we're programmed to carry out in the pursuit of friendship, let's indulge a high-minded version of how we choose friends, and best friends in particular. (And while it's more heartwarming, this view is ultimately not incompatible with those other theories; it just looks at friends from a different angle.) Let me guess:

The wide-awake side of your brain is not contemplating your genes and how crucial it is that they be passed on, nor is it constantly planning how you'll survive in an emergency. Instead, it's preoccupied with questions of who you are, who you want to be, and what you need (and want) to get done in the short and long term. Your identity and your goals dominate your thoughts, and finding friends who can reinforce both of those is productive, rewarding, and very helpful.

Psychologist Carolyn Weisz, Ph.D., of the University of Puget Sound, has found evidence that best friendship offers "social identity support" or an affirmation of one's sense of self and place in a desired group. At first glance, this might seem like another way of saying that people like people who are similar to them. But here "identity" goes beyond liking the same things, and "support" means more than touchy-feely cheerleading, says Weisz. "Your self-esteem doesn't just come from your individual efforts. It also rests on the group to which you belong and the status of that group."

Say Anna wants to be an actress. Her friend Eva is an actress who already views Anna as a fellow thespian and buys into her friend's dream, confirming the way Anna sees herself. If Eva invites Anna to be in her drama club, Anna's identity and self-esteem are simultaneously strengthened. And then, if the drama club puts on an award-winning show, Anna's pride in both herself and the group will skyrocket. Eva will share in this pride, and her very presence will remind Anna of the forward movement they are making toward their shared goal. Anna will be flattered at the notion that the outside world perceives her and Eva as two peas in a pod. Aside from emotional and internal benefits, Anna will receive practical help from Eva and from the drama club to which they both belong. She'll get feedback on her acting, which she can use to improve, and tips and connections that could help her land other roles.

Or take the real-life example of Solomon Dumas, twenty-two, and Slim Mello, twenty-five, the two closest friends within the Ailey II dance ensemble—renowned as the second company of the famous Alvin Ailey American Dance Theater and a launching pad for young performers' professional careers.

"I can't describe how exactly we became really close," says Solomon. "When you click, the grooves just fit. You can't force a friendship. But we're both a little older than a lot of the students at Ailey. So we're a little more serious. I think we gravitated toward each other for that reason."

Slim came to New York from his home country of Brazil in the summer of 2009. Ever since his dance teacher showed him a video of an Ailey performance years ago, his single goal has been to be in the company. "Every day is a blessing to me, because I'm close to my dream," he says. "We work hard, but working is good," Slim says. Solomon, who hails from Chicago, began formal dance training at the Chicago Academy for the Arts. He recalls how his mother always supported his ambitions, setting aside what little additional money there was for arts camps or classes.

Ailey II members take classes in the morning and rehearse until six. Before hitting the road for tours, they might learn ten ballets in a month, sometimes mastering two roles for each piece. "There's lots of sweating!" Slim says.

Keeping up perfectly with the grueling schedule and intricate steps still doesn't guarantee sustained success in such a competitive field, where hundreds of capable people fight for a few paying spots. That's why Slim and Solomon help each other through the physical and the spiritual trials of their vocation. "No one can understand better what I go through than Slim," says Solomon. "He's right there with me. We know exactly what the other is feeling."

"Solomon is so focused," says Slim. "All the time we've been onstage together and the time we spend talking on the road have made me grow a lot, professionally and personally. And when he speaks about dance, he can really move you. He's so mature. I wish I could speak like him." The previous year, Slim was feeling down around his birthday, since it's normally a day he spends celebrating wildly with his friends and family in Brazil. He hadn't seen them in two full years. Solomon remembered and insisted on taking him out for drinks. "I felt really good," Slim says. "I didn't have my family, but here was someone who really cares about me. He is my bro. Here in the U.S., I have my job with the company, I have my girlfriend, and I have Solomon."

Though Slim takes Solomon for a role model, the feeling is mutual. "Slim came to this country for one reason," Solomon says. "At times I've taken what I'm doing for granted. But to see someone come here by himself, without knowing the language—that inspires me. It helps me to stay focused." Slim also impresses people with his unfailingly generous attitude toward his fellow dancers. "When you're on tour with the same people six months out of the year, you can get frustrated with each other. It's a lot of pressure, it's a lot of competition, and we all want the same things. But even when situations don't seem fair, I've never heard Slim complain to anyone or talk disrespectfully about anyone. That inspires me to not be petty and to become more aware in my interactions with others," Solomon says. "Slim has that classic dancer's build, and amazing feet, but it's what is inside him that makes him so compelling onstage."

Best friends don't have to share an identity per se, but they do need to support the other's view of himself and make each other feel great about their pursuits. Weisz asked a group of college freshmen about their close friends and used questionnaires to determine whether they received social identity support from them. She then followed up five years later, when the students had graduated and moved off campus. Social identity support didn't predict whether the friendships generally endured, but it did predict whether one of the friends became a best friend. Part of maintaining a close friendship, Weisz points out, is supporting someone's identity as it inevitably shifts over time. "When one friend gets into exercise and healthy eating, it can be hard for her to not become judgmental of a friend who doesn't have a strong desire to get in shape," she says. "Or when one friend comes into money and wants to take nice trips while the other can't afford them, these changes are difficult to work through."

The Shapes and Sizes of Friend Networks

If we move our camera even higher up to an aerial view of how friendships are formed, we can see some interesting patterns. The British anthropologist Robin Dunbar, Ph.D., discovered that the size of a primate's brain is correlated with the size of the social group within which its species typically lives. The magic number for humans—extrapolated from our average brain size—is 150.

More specifically, Dunbar conceives of the number 150 as embedded with a number of layers. "In effect we have five intimate friends. Fifteen close friends, 50 good friends, 150 friends," Dunbar says. "The 15 layer has long been known in social psychology as the 'sympathy group' (those whose death tomorrow would seriously upset you). Beyond 150, we have acquaintances, and here they are more often asymmetric (I know who you are, but you don't necessarily know who I am). The 1,500 layer seems to equate to the number of faces we can put names to."

Some individuals might be particularly equipped for large groups of pals on the basis of the size of a certain area in their noggins. In one study, Dunbar asked subjects to write down the names of everyone they had been in touch with (socially, not professionally) over the previous week and then measured the volume of their orbital prefrontal cortex with a magnetic resonance imaging scanner. He also gave them tests of "mentalizing," measuring how able they were to comprehend others' states of minds while interacting with them—an important social skill. It turns out that the size of this part of the brain correlates with the size of the person's social circle, as measured by the number of names on the person's lists, as well as with their mentalizing ability, thought to be centered there.

If bigger is better, it makes sense that more empathetic people would have more close friends. But another study of Dunbar's found a negative relationship between the average closeness of one's friends and the size of one's social network. There is apparently a limit to how much emotional intensity one person can sustain with a large group. Just as the friend niche limitation model suggests, a balance of quality and quantity marks the most satisfying setup.

Another team more recently found a correlation between the size of the amygdala, a brain region that processes emotional stimuli, and both the size and the complexity of a person's social network. Like Dunbar, this team wanted to know how many people each subject was in touch with regularly. Additionally, it identified a group affiliation for each contact, such as an amateur basketball teammate or a childhood pal. If a person's contacts came from different groups, her network was deemed more complex than that of someone whose friends were all from one group, making for a simple social life, as was the case for the characters on *Friends*. But while those with bigger amygdalae had larger and more complex social networks, again, the size of the amygdala was not related to how happy subjects were or even how supported they felt in life. It just seems to enable (or at least grow in response to) a sizable and disparate gang.

It's worth pondering the finding that network complexity doesn't lead to happiness, because our networks are getting more complex. Back when we developed our capacity for 150 contacts, we didn't have much choice over our group. The ability to pluck out friends from different corners of our lives seems like an advance, but Dunbar doubts it:

> If anything, it may be a disadvantage. In traditional societies, everyone's 150 overlaps with everyone else's, so everyone knows everyone else. That gives the community a great degree of structural strength, in particular through internal peer-pressure self-policing. What has happened with modern, post–World War II economic mobility has been that our social networks have become fragmented and geographically dispersed. The result is that they consist of a set of smaller sub-networks that don't overlap all that much (family, college friends, friends from the first job in the Far East, friends from the current job in London, etc.). As a result, these networks are less

> dense, less interconnected, and so we can maintain different personae to different groups. That gives us more freedom, certainly. The cost we pay is that networks are less mutually supporting.

If all of your friends know one another, you must present a consistent self to them, or at least, they will ascribe one to you. People in a clique know what is going on with each person and share information efficiently, as Dunbar points out. If you have close friends in several different circles, you have to present yourself and anything you're going through multiple times. You might tailor your personal narratives to each of these people differently and feel a bit fragmented as a result.

A friend once commented to me that she admires that I have some friends who are quite different from me. I appreciated the compliment, on one hand—I always prided myself on being able to connect with foreigners, for example, since I enjoy learning about other cultures—but I pointed out to my friend that while she has a more interconnected and homogeneous group than I do, I felt that reflected her strong sense of self. She knows exactly what she is interested in and what she believes and has friends who reflect all that effortlessly. I, on the other hand, sometimes find myself slipping into each of my friends' worlds, chameleonlike, meaning all of my sides are rarely expressed at once.

What's more, our friend networks are remarkably unstable themselves. A study by a Dutch sociologist who tracked about a thousand people of all ages found that on average, we lose half of our close network members every seven years. To think that half of the people currently on your "most dialed" list will fade out of your life in less than a decade is frightening indeed.

Why, if making true friends is so hard, so rewarding, and so endemic to our emotional and cognitive wiring, do we let so many slip away?

Friends Change the World (or Your World)

Sofia and I met in the dorm—that petri dish setting where close proximity to others makes it hard not to meet people, and where inherent demographic similarity and potentially resonant burgeoning identities give a few hallway chats a fair shot at expanding into a lifelong relationship. Our meeting is important to the two of us, but such encounters can occasionally trigger events that affect the masses. Sergey Brin and Larry Page also met on a college campus, for instance. The year was 1995, and Brin, a computer Ph.D. candidate at Stanford, was assigned to show Page, a prospective student, around. If a friendship hadn't formed out of that encounter, it's very possible that we would not have Google, meaning, perhaps most devastatingly, that you never would have been able to google your old friends.

Some historians argue that the friendship between President Franklin D. Roosevelt and Prime Minister Winston Churchill, though certainly not solely responsible for the Allied victory in World War II, did substantially smooth over circumstances that could have made cooperation far more difficult and diplomatic outcomes less fruitful. The FDR Presidential Library and Museum refers to the friendship as "one of the most extraordinary relationships in political history, a relationship marked by an intimate correspondence unparalleled among national leaders, a relationship which, in due course, would lead to the establishment of a military alliance unique among sovereign states." The library cites the creation of the United Nations as a concrete result of the high level of harmony and respect between the two friends.

Just as friendship can be a good and a bad influence in our little lives, so it goes on the global stage. Albert Speer, an architect who delighted Hitler with grandiose plans for the German skyline, is often referred to as Hitler's best friend. Speer didn't make the Nazi leader evil, but his fervent devotion to him did advance the cause substantially. Hitler put him in charge of the German war machine despite his lack of formal expertise, and being naturally gifted at the task, he ramped up tank production fivefold and plane production fourfold before the war ended.

EVEN WITH ALL THE INSIGHT we now have into why we're driven to make friends and why we become close to some and not others, the chances of any particular friendship's gelling and enduring can't fully be predicted. Whether the development of these bonds is preordained or a random happenstance is really yours to decide, according to your own philosophy of life. You'll never be able to completely account for all 150 of your modern tribal ties. But now you're perhaps primed to notice those everyday moments that could transform a nameless face into a dear one with a name that is etched into your soul, or at least programmed into your phone.

CHAPTER 3

Friendship in Childhood: Pals and Nemeses

"SHE WAS SO HAPPY," RECALLS Suzanne Ludlum of her best friend, Denise, whom she met in kindergarten. "I remember seeing her on the playground, with her white-blond hair. She always smiled at me, and her whole face would light up." Suzanne and Denise lived in the same town, Denville, New Jersey, but had to cross a field and a stream to get to each other's houses. Nevertheless, "we were inseparable," says Suzanne, now fifty-two.

Tragedy befell Suzanne in her twenties: Both her parents and her brother passed away during that decade. "Denise is the one who is there when I need a 'family member' to talk to," Suzanne says. "I've often felt like an orphan. She is the only person in my life today who knew my family, and it helps me so much to know that she remembers them as I do."

Suzanne and Denise both got married, and Suzanne had a daughter, now twenty-seven. "Denise's husband works for the government, so they moved around a lot," Suzanne says. "I've visited her in most of the places she has lived." For a time they grew apart, but once Denise had her sons, the two became close again, bonding over motherhood. Suzanne, now a yoga teacher, got divorced and remarried, and had another daughter, who is now a preteen.

Four years ago, Suzanne, who lives in Virginia, had to have surgery on her spine. Denise flew from her home in Florida to take care of her friend after the operation. A few weeks before I spoke to her, Suzanne had traveled to Florida to watch Denise's children while Denise went to Cleveland, where her husband underwent a heart procedure. "We call each other in the middle of the night," says Suzanne. "We're each other's sounding boards." In a touching display of empathy, Suzanne was worried about Denise's ability to handle her husband's illness. "This must be so devastating for her because she's never had anything like this happen to her," she said as her voice cracked. "I've had a lot of loss, but she hasn't, so I think her pain must be great right now."

Many childhood friendships dissolve, leaving behind just a few fuzzy memories; others, like that of Suzanne and Denise, lend a steady beat of continuity to life. Whether or not you're still in touch with your old pals—or can even recall them clearly—they surely helped shape you, for better or for worse. In fact, the qualities that enable childhood friendships are the same ones required for general success in adulthood. As prominent child developmental psychologist Kenneth Rubin, Ph.D., of the University of Maryland, puts it, "The better able children are to form good, sustaining friendships and to be accepted and valued within their peer groups, the more apt they are to do well in school—and, in the long run, in life."

Attached to Parents, Fascinated by Friends

Peter Gray, Ph.D., a professor of psychology at Boston University, once asked whom children want to please, their parents or their peers. The answer? Their peers. Said Gray: "Are they wearing the kind of clothing that other kids are wearing or the kind that their parents are wearing? If the other kids are speaking another way, whose language are they going to learn? And, from an evolutionary perspective, whom should they be paying attention to? Their parents—the members of the previous generation—or their peers, who will be their future mates and future collaborators?"

Gray brings us down to a child's point of view, the one that inspired those comically invisible and unintelligible adults of Charlie Brown's world. Kids' movies and books in general are manifestations of the friends-focused mind-set. The Harry Potter series, for instance, is primarily about schoolmates and the three-ring friendship of Harry, Ron, and Hermione. "In children's literature, the character often has adventures apart from the family and is sustained by friends along the way," says Philip Nel, a professor in the Department of English at Kansas State. "It's a common motif that reflects the transition from family to being off on one's own, forging new relationships."

Anatomy of Childhood Chums

Friendships sprout much earlier than you might think. A one-year-old who has the chance to interact regularly with other little ones will indeed choose favorite playmates—first friends. Toddler buddies frolic in more complex ways than do non-friends. They might engage in pretend play, such as acting out "mom and baby," which requires more cognitive skills than tag or other literal pursuits. In one study, the first time any of the children being observed switched over to symbolic play was always with a friend rather than with an acquaintance.

Toddlers who entered a new childcare center with friends in tow adjusted better than those who moved to a place where they had no pals. And while most in the clutches of the terrible twos or threes ignore other children's cries (they are usually too self-centered to be upset by another's distress, even if the bawler is right in front of them), they are three times more likely to tell the teacher or give comfort when a friend bursts into tears than when a nonfriend is melting down nearby.

Little kids are on the lookout for playmates who enjoy the same fun things they like to do and who are pretty easy to get along with. Starting around age ten, though, a friend is not just a convenient companion with nice toys, but is someone who is recognized for his or her values—such as the ability to stand up for others—as Rubin points out in his book for parents, *The Friendship Factor*.

Once they reach the age of eleven or twelve, kids expect a measure of concern and thoughtfulness from their pals. Bonds aren't just based on shared experiences anymore; they're sealed with emotional revelations and repeated instances of "getting" one another.

Psychologist John Gottman, Ph.D. (now famous for his supposed divorce-predicting abilities), charted the elements of friendship formation and upkeep among kids: They talk and seem "connected" emotionally, they tell each other bits of information about themselves, they figure out what they have in common (such as love of the same activity), they start to open up about their feelings, they pick up and run with it when the other begins a game, and they try to work out conflicts or disagreements. Kids may exchange fairly in the beginning of their relationships but soon move to a more pleasant "communal orientation," giving what the other needs, not what the other is owed.

Luckily, most kids have at least one "mutually agreed-upon" friend; they are part of a two-way, voluntary relationship. But children's friendships are not particularly stable. A study from the '80s found that 31 percent of fourth graders did not hang on to a best friend for six straight months; another found that 67 percent of fifth graders didn't keep their three closest friends after a year's time.

Just like grown-ups, kids are drawn to friends who are similar to them, especially in terms of gender, race, age, and interests. Sharing a socioeconomic status is more important to teens than to young kids, though, and Rubin did find that if children are in a racially diverse environment, they will choose friends on the basis of behavioral similarity, not skin color.

But it could be that similarities in values, interests, and perceptions, which increase over the course of a friendship, are the result of that friendship, not its cause. How exactly do little pals become more alike? It seems to happen via conversations and even pretend play, wherein mutually created characterizations and plots shape kindred worldviews.

Drew, a sunny eight-year-old boy from Moorestown, New Jersey, says he is best friends with Justin because "he's fun to play with and he's nice to me." The two boys are obsessed with PlayStation video games, were members of the same local baseball team for years, and shared a teacher back in first grade. "I met Justin when I was six months old," Drew says, demonstrating how friendship between mothers and fathers often enables bonds between young companions. "If you meet a friend when you're little," Drew concludes, "you'll probably be friends for a long time." Of course, mutual affection must be present, in addition to common passions.

While there are exceptions, boys and girls remain segregated in most schools and playgrounds. When boys were asked about important activities, they ranked playing sports as number one. Though the girls surveyed spent almost as much time playing sports (albeit noncontact kinds) as the boys, the most important activity for girls was talking to their friends. Indeed, girls have more intense one-on-one friendships at an earlier age than do boys, who are more likely to run in groups.

When I asked Ella, a thoughtful ten-year-old from Bethesda, Maryland, why the boys and girls at her French immersion school don't tend to fraternize, she sighed and said, "I don't know. I guess it's because the boys think they're so cool." Though Ella is herself part of a local championship soccer team, she explained that while most of the guys spend recess playing soccer, she hangs out with her small group of close girlfriends. As with other rules of childhood friendships, crossing the gender line is something kids seem to subtly discourage each other from doing, perhaps because mixing roles and expectations threatens their nascent identities.

It's hard to gauge the impact of one kind of companion, the sort that exists only in kids' imaginations. I had two grown men, who often argued with each other, for imaginary friends as a little girl. I don't think they affected me too much, though they must have convinced my parents of a certain strangeness in their daughter. Yet the fact that I counted figments of my mind as pals wasn't so weird; in the United States and Europe, almost half of all kids will have an imaginary friend at some point. Funnily enough, even though children often know these friends aren't "real," they claim to get just as much help and affection from them—and experience as much conflict with them—as they do with real pals.

Amiable (and Otherwise) Personalities

Before a kid can be influenced by her friends, she needs to make them—or risk being negatively influenced by a *lack* of friends. How she is parented does matter, but both how she is treated by her parents and how she interacts with people outside her home are molded by something she's (at least partly) born with: her temperament.

So much for the blank slate. Temperament reveals itself very early on. "Before a child is walking and talking . . . we can begin to understand, for instance, whether she is likely to be outgoing and confident around other children, or, in contrast, cautious, worried, and wary," writes Rubin. He outlines three behavioral tendencies of kids: to move toward other children, to move against other kids in an uninhibited way, and to move away from them, out of fear or other negative emotions. These orientations are dubbed normal, aggressive, and withdrawn, and they often predict a kid's pattern of making friends.

What makes a kid likable? It's unfair, but looks matter; attractive children are especially likable. It helps to have a name other children find appealing, too. And in accordance with the rigid kid rule of gender segregation, young people like boys who play as boys "should" (girls are allowed a little more role flexibility). Likable children size up new situations on the playground or in the classroom and fit themselves into the scene. They are assertive, but not hostile. They don't talk about themselves *too* much, and they communicate clearly. They're kind, helpful, generous, and thoughtful. They're warm and pleasant and tend to have a sense of humor. Unsurprisingly, they're able to make and maintain friendships.

Children who are natural social stars, Rubin adds, present themselves "successfully to others by putting on somewhat different faces for different audiences. . . . They understand when to put on which face, without ever appearing shallow or false to others and without feeling like fakes or frauds. In short, these are children who are sensitive and responsive to social cues." This is the child who knows how to work the room with jokes or dance moves at her own birthday party with her adoring relatives, but who also knows to hang back and let a friend shine at *his* birthday party. (This talent for decoding others and adjusting one's own behaviors and expressions appropriately is lacking, to one degree or another, in children on the autism spectrum and is why they often have a particularly tough time relating to other kids and forming friendships.)

Aggressive kids, or those who naturally "go against" their peers, have several disadvantages in the friendship market. They tend to be unable to appreciate the thoughts and feelings of fellow children; they react before thinking (especially if they've been rebuffed by a potential playmate), and they misunderstand others' intentions, often reading the worst motivation into a benign interaction.

An anxious or withdrawn child, who "moves away" from peers, is easily overwhelmed at the center of the social universe, Rubin says. "He'll back up and back up until he bumps into another socially anxious kid. These two can be supportive of each other after a fashion, but not much fun. It's not the sort of lively interaction that two nerds could have." Shy kids are more likely to have shy friends (though that doesn't mean all of them pair off thusly), and aggressive kids, if they can make friends at all, are more likely to have aggressive friends. Here at the nexus of personality and friendship is the origin of a difficult dynamic often called the Matthew effect, after the Bible verse declaring that those who have will get more and those who don't will lose what little they do have. Shy kids influenced by shy friends are pushed deeper into their limited comfort zones, and forceful kids with brutish buddies are positively reinforced for impulsive (and repulsive) behavior.

Another metaphor that captures the relationship between temperament and friendship is the classic chicken and egg quandary. Is the friendless child withdrawn because he blames himself for his solitude on the playground, or is he friendless because his shyness prevents him from reaching out to other kids and even accepting small advances from them? The ramifications of this cycle can be long term: Some children who are socially isolated have been shown to develop depression that can "snowball" quickly into a serious condition in later years. This kind of depression can be stalled in kids who were isolated but who then developed friendships.

If the personality divide is breached by kids themselves, perhaps with a touch of encouragement from parents and teachers, the shy child can befriend a sensitive extrovert who coaxes her into new situations, and the aggressive kid can attach himself to a patient soul who teaches emotional regulation by example. Finding a best friend who is socially competent, Rubin writes, is an effective way for formerly "unlikable" kids to get off the downward path and onto an uphill route.

It's difficult to convince a child that she should reach out to friendless kids, since the cold reality is that unless she is extremely empathetic, she'll intuit that the friendship won't do much for her and could even harm her social standing. Parents of socially gifted kids should push them to connect with outsiders, not just to prevent the extreme cases of bullying we sometimes hear about, but to counteract the heartbreaking loneliness of a friendless kid, whether or not he's victimized. A Canadian study of kids with physical and developmental disabilities by Anne Snowdon, Ph.D., of the Odette School of Business at the University of Windsor in Ontario, found that 53 percent of disabled kids have just one friend or no friends at all, and just 1 percent spend an hour a day with friends outside of school. Even if they don't need more friends themselves, kids can learn lessons in reaching out and giving by drawing outcasts into their circles.

Those Who Have, and Those Who Have Not

When Drew was four, he pronounced his *R*'s perfectly. But when his preschool pals came over, his mother noticed that "Ride my bike" became "Wide my bike." Why the sudden linguistic lapse? His little friends couldn't say *their R*'s yet, and he naturally adjusted himself to their manner of speaking. (Relatedly, a big developmental clue to the primacy of the peer group—alluded to by Peter Gray—is the fact that a child who grows up in a house of immigrants will develop the native accent of her peers, not the foreign one with which her parents speak.)

A caveat is in order, however: There is still much we don't know about the long-term developmental consequences of friendships that begin in early childhood. Most studies are school based, leaving a dearth of information about friends who aren't classmates. And many are based on self-reported rankings or nominations of friends by kids or observations of parents and teachers. Watching kids interact naturally, over time, is of course much more difficult for researchers to pull off.

Regardless of the work that developmental psychologists would like to do to add to our understanding of friendship, the "outcomes" they use to measure the effects of friends are somewhat limited and general. Even if your childhood friends didn't alter your major life choices or ultimate level of success—at least not in obvious ways—they added colors and textures to your makeup that are impossible to capture in a lab but significant to your life story, the narrative that defines you and helps you make sense of your time here on earth. That's why personal stories of the influence of friends (an influence that is always interwoven with concurrent influences from your family, romantic relationships, circumstances, etc.) are as powerful as the litany of proven benefits of friendship.

As for kids, it's a good thing that the majority of them have at least one friend, because a big-picture conclusion of researchers is that friendship is essential to social and emotional development. Via conversations with pals, where they learn each other's stories, dilemmas, and thoughts, friendship strengthens kids' perspective-taking and moral-reasoning skills. Caring, high-quality friendships boost all-around psychological well-being.

You may recall that around ages nine to twelve you and your friends were able to give each other practical help, advice, lively companionship, validation of hopes and fears, and a chance to share intimate secrets and feelings. Friends at this stage offer a model for later romantic and parental relationships, while at the same time providing an "extra-familial safe haven," as Rubin and his colleagues call it.

When she is disagreeing with a friend, a child evaluates the friend's judgments more positively and is more willing to back down from her own position than she would be if she were at odds with a mere acquaintance's stance. Since friendship gives kids the security to alter their points of view, it also sharpens their critical thinking skills.

I spoke to five gregarious girls from a suburb of Columbia, South Carolina, all neighbors, between the ages of eight and ten, who play together regularly. In between laugh attacks over their private jokes, they casually demonstrated the ancillary lessons of friendship. "We share secrets," said Ashley, eight, demonstrating how the same bonding processes adults participate in—revealing feelings and gossiping about others' private affairs—start in childhood. Grace, a ten-year-old, added, "But you can't trust some friends. They might tell other people." I asked Grace how she knows whom she can trust. "It's trial and error," she answered. "Sometimes you learn the hard way. And you don't want someone who will judge you on your secret—like if you get a bad grade on a social studies test, you don't want them to think badly of you. But when you find one you know you can trust, it's fun and cool to share secrets and to know something that no one else knows."

As friendships are embedded within a larger group governed by laws (e.g., jeans with the green triangle on the back pocket are bad; jeans with the red triangle on the back pocket are good), pals can offer insightful interpretations of the byzantine rules and gentle explications to friends who don't seem to be getting the memos.

Fitting in is especially difficult for kids whose backgrounds are different from their schoolmates', and that is part of the reason that Astrid, now thirty-five, treasures her childhood friendship with Sara, a fellow Iranian Armenian immigrant. "Sara moved to my town in New Jersey when we were in fourth grade. I had already been in the U.S. for seven years by then; consequently I felt responsible for her," Astrid recalls. "Sara became a kind of life preserver against the very intense shame and discomfort of being an 'ethnic' kid. I had my buddies and was not blatantly some sort of outcast. But I was totally mortified by my parents' accents and general foreignness, and they were quite conservative and old-country about certain things.

"Sara had an early mustache and unibrow and was way darker and hairier even than I was," Astrid says. "I certainly recognized that she wasn't exactly raising my stock, but my stock wasn't very high to begin with." Soon Astrid took charge of Sara's adjustment by forcing her to practice for cheerleading tryouts for the peewee football squad. "All the girls would gather at recess and practice the cheers, and later I would take Sara aside and have her repeat them over and over again. She sounded totally ridiculous with her accent and couldn't get the hang of the words, which made no sense to her. Neither of our parents were actually going to allow us to try out for cheerleading—the skirts were too short—but I was desperately trying to make her fit in better, and I probably gained some confidence for myself in the process. Like, there but for the grace of a green card lottery go I."

Since Astrid's parents implicitly trusted Sara's, the girls were allowed to pass a lot of time together, enabling their friendship to deepen. "By having Sara in my life, I was able to spend some hours of my days not feeling stressed out about seeming different. I could talk about family things without trying to pretend my life was different than what it was." In contrast, Astrid's ethnicity was thrown into relief in the company of her all-American classmates; ironically the time she spent with Sara made her both forget her "differentness" yet also embrace her identity, in a long-lasting way. "Neither of my siblings nor Sara's brother had a close Iranian or Armenian friend at school, and now all of them are very distant from their ethnic identity, whereas Sara and I are both still quite involved with our respective communities," Astrid notes. "To me, finding ways to bridge this bicultural experience is about being able to keep yourself whole, which is why I am grateful for having had a friend who supported that bridging."

Best friends do seem to have a special power; some researchers see intimate friendships as intense training grounds that school kids in conflict resolution and empathy. That's why a trend among some parents and educators to discourage best friendships—out of an understandable fear that besties won't then branch out and include all available playmates—is troubling to many psychologists. Any mission to pull bosom buddies apart is probably fruitless anyway, given their prevalence: A Harris Interactive survey of Americans ages eight to twenty-four revealed that 94 percent had a close friend.

Ella, the ten-year-old from Maryland, has two best friends, Maya and Sophia. With a touch of embarrassment mixed with pride, she told me that together they had created a clothing company. "Maya was over yesterday and we made a dress. Well, we went to a thrift shop and bought a dress, and then we changed it." (She beamed when I told her that many Hollywood stylists charge thousands of dollars for similar endeavors.) When I asked how her friends have changed her, she said they've made her a bit more "girlie." The threesome loves to have sleepovers and "make up stories about our future." A few days earlier, they had purchased identical rings at the mall. "We decided that since we're best friends, we needed something to show it. I felt even closer to them after we got them. The rings say, 'I love my BFF.'"

Ella assumes she'll be close to Maya and Sophia forever; we adults know that's sadly not likely, yet the obvious happiness and inspiration the girls give one another now will leave an indelible, positive mark.

At one end of the scale, kids like Ella with close, warm friendships flourish and learn to be better, more realized human beings. At the other end, those without any friends at all can feel terrible pain. Friendless children are lonelier than other children, unsurprisingly, and—back to the chicken-and-egg conundrum—they tend to be socially timid and sensitive. They are also likely to grow up to be anxious or worried adults. Though the thought of an innocent child sitting alone at lunch pulls the heartstrings, such children are often (again, probably because of both their temperament *and* their consequent friendlessness) less altruistic and positive in their interactions with peers. Aggressive kids—often disliked by classmates and teachers alike—are targets for rejection, just as are timid children.

But even kids who have been actively rejected by their classmates en masse can find legitimate psychological protection in just one friend. (A younger kid whose maturity might match that of the cast-off child's is a good candidate—especially if he doesn't attend the same school and is indifferent to that institution's position toward his older buddy.) Rejected kids with a friend, for instance, are more trusting than those without one.

Familial Forces on Friendfluence

The debate over the relative contribution of parents, friends, peers, and genes to a kid's fate continues, but some particular pieces of this ultimately unsolvable puzzle are worth looking at. Preschoolers who are securely attached to their mothers enjoy being with friends and cooperate with them more than those who are not securely attached. Affectionate, approving, and encouraging parents tend to have kids who are friendly with their peers. On the flip side, a lack of parental warmth is associated with socially maladjusted children. Summoning Goldilocks, researchers found that preschoolers whose parents are overstimulating are more likely than others to face rejection. Of course other factors, such as whether the child has space to invite other kids over and the child's own personality, which can solicit or discourage parental openness, are also in play.

Divorce is linked to social incompetence in kids, though the relocations and economic instability associated with divorce, not the breakup per se, are likely what cause social problems. If you're a parent who likes to socialize with your friends but fears it takes away from quality time with your brood, take heed: Moms and dads with friends whom they see regularly tend to have well-adjusted kids with healthy friendship networks. Proof that the aforementioned puzzle can't be completed: These parents may have naturally good social skills, meaning that even if they stopped seeing their comrades regularly, their kids might still learn (or possibly even inherit, to an extent) what it takes to be well liked.

Friends trump siblings in kids' hearts: Five-year-olds show more affection and reciprocity to their pals than to their little sisters or brothers. And parents thinking of a second child have in their midst an important clue as to what kind of sibling their firstborn will be—whether or not he is a good friend. Researchers watched a group of four-year-olds interacting with their friends prior to a younger sibling's birth. How positive the preschoolers were with their playmates predicted the quality of their relationship with their little siblings as teenagers. Kids who are good friends to other kids are likely to be nice to a new member of their family.

The Classroom, Up Close

If you were like most children, you had friends growing up. But you and your friends were not alone. You were part of a peer group with roles that seemingly must be filled no matter which individuals find themselves in the same social ecosystem because of geography and birthdates.

What you might call archetypes or recurring characters of every movie portraying a schoolyard (the snotty popular girl, the jock, the smart girl next door, the rebel, the victim, and the invisible kid) are part of what developmental psychologists call "sociometrics." John Coie, Ph.D., of Duke University, and colleagues created a typology and tested it against the composition of American classrooms. "Popular" children comprise about 15 percent of kids. "Accepted" ones are your average Joes and Janes: They have friends, even if they're not wildly liked by a majority. They make up a full 45 percent.

"Rejected" children are subdivided into "rejected-submissive" (the shy, anxious ones) and "rejected-aggressive" (the over-bearing ones with tempers). Together, they account for about 10 percent of the average student body. Rejected kids don't fare well in life, especially if they are rejected over the course of several years. Now that tech gurus rule our world, a real "revenge of the nerds" has transpired, reinforcing the notion that cafeteria fortunes can be reversed. But whereas geeks often have geek friends, rejected kids are usually friendless or repeatedly thwarted in attempts to sustain bonds with peers. There's a big difference between being part of a tight chess club that gets harassed by the football players and being a bullied victim who is part of no club at all. The "I'll show them" fantasies of rejected kids, according to the research that has been done, are sadly unlikely to be realized.

Four percent of kids in Coie's typology are "neglected." No one picks on them, but no one really notices them either. Teachers find them well behaved and sometimes aren't aware of their isolation. Finally, the intriguing "controversial" children also weigh in at 4 percent. These kids are adored by some classmates and loathed by others. Think class clown or leader of the rebels. (A few children in any of these assessments fall into the "ambiguous" category.)

Like animals giving each other chemical signals to govern herd behavior, no one kid declares who will be the rejected one in his or her class. "It is done through thousands of little incidents, each followed by a gossipy discussion: 'Did you see him do that?' 'Did you hear what she said?' 'I can't believe she did that!' " writes Michael Thompson, Ph.D., coauthor of *Best Friends, Worst Enemies*. "Whether verbally or silently, a group consensus is reached that the child is dangerous or excessively weird." Once someone is declared an outcast, the winds of perception forever knock him down: Children are five times more likely to assign negative intentions to the actions of a rejected child in comparison to the same behavior carried out by someone in the popular category.

Meanwhile, some kids might literally be born for popularity. In their book *Connected*, James Fowler, Ph.D., and Nicholas Christakis, M.D., report that in one study, genetic factors accounted for approximately 46 percent of the variation in how popular kids were. "On average," they write, "a person with, say, five friends, has a different genetic makeup than a person with one friend." It's possible that the genes in play here relate to temperament, and different temperaments require different numbers of friends for social satisfaction.

"Everyone at our school basically has friends," says Grace, one of the preteens from South Carolina. "But we used to have a kid named Joshua," Ashley says. "He didn't know a single thing. I think he was homeschooled. We tried to make friends with him, but he wouldn't talk to us. He was a little bit mental, and one time he didn't take his medicine and he was running around. The teacher had to send him to the office five times! Then one day he left and everybody was like, 'Yes! He's out of our lives forever!'" Joshua was the chosen reject in Ashley's class. With his behavioral issues he was clearly a prime candidate. That the children listened to the better angels of their nature and tried to connect with him didn't alter his designated role in the end.

That's why James Olsen, Ph.D., of the University of Memphis, thinks (as apparently Joshua's parents thought) that removing a rejected child from a school is a solid plan. "It's very hard for a kid in that situation to climb out," Olsen says. "Removing a kid is not a perfect solution because there's a reason he got there in the first place, but it's a good step. It would be wise to also beef up his social skills. Something like learning karate could help, because it teaches emotion regulation." Another solution of sorts for a kid like Joshua who is at the bottom of the heap is to somehow befriend a cool kid. "It will raise his profile," Olsen says, "but it will also diminish the popular child's status." Beware, the ecosystem *will* be balanced.

Friendship is different from popularity, of course, but the two are related, in that popular kids do have more friends. Having a friend melts away loneliness at all levels of popularity (some queen bees feel adored, perhaps, but not truly known or understood). Rubin makes a distinction between "popularity-as-decency"—kids who are well liked because they are kind and fun—and "popularity-as-dominance"—those who are revered because they are good-looking, wear the right clothes, or are captains of teams. "I would wish popularity-as-decency on my own kids," says Rubin. "And I'd choose friendship over popularity any day." Of course it's easier for an adult, who is not as bound to a set peer group as a child is, to say such things.

In fact, having once been shut out from the popular clique is a status-enhancing badge of honor for many grown-ups. In her book *The Geeks Shall Inherit the Earth*, journalist Alexandra Robbins makes the case that kids who are excluded for strange interests or a lack of a desire to fit in are celebrated for those same traits once they become adults. She calls it "quirk theory." I think it's a nice (though sometimes falsely modest) way for smart, talented, yet quirky adults to reframe any bad experiences they had growing up, but I still suspect that being left out is not so pride-inducing for the young boy eating alone in the cafeteria, even if he expects a brighter future.

Grace and her friends seem attuned to "popularity-as-dominance" but hip to the value of "popularity-as-decency," too. "Clothes are important. You don't want to look too fancy, but you also don't want to wear old, ugly clothes," says Ashley. "There are friends who won't judge you for your clothing, but then there are friends who are like, '*What* are you wearing today?' "

"And then there's the electronics," says Grace. "If you don't have a Nintendo DS, you can't be in this group, or if you don't have an iPod, you can't be in that group. Or sometimes, for a group that is trying to grow up a little fast, you have to have a crush or you can't be their friend. Oh, and if you wear makeup, you're automatically popular." When I asked them if it is more important to have a friend you care about or to be popular, Grace jumped in: "She could be a nerd and I could care less, as long as she's nice to me, and we're friends."

Children are socialized in a group but ideally gravitate toward friends who help them manage the group while also offering a respite from its limited roles and codes. As Thompson puts it, "Groups are the highways of childhood. Our kids are swept along, going at the same speed as the majority of the traffic. . . . Friendship, by contrast, resembles the side streets and back roads of childhood. Friends can go at their own pace; they can stop when they want to; they can get away from the speeding traffic."

Many factors converge to create episodes of childhood friendship drama that can still make adults cringe years later. When Christine, now thirty-four, was finishing up the fifth grade at a new elementary school in California, she threw a pool party in her backyard. She invited every girl in her class except one, Margo. "Margo was very popular, and I would have been thrilled to have her, but I knew she was going to leave for camp right before my party," says Christine. "Unfortunately I also made the mistake of inviting Melissa, a friend of Margo's, who was going to camp with her. So when Margo heard about the party, my reasoning—that she wouldn't be able to come because she was at camp—would not have been obvious because that didn't stop me from inviting Melissa. I took it for granted that Margo knew that I hadn't meant to exclude her. It never would have occurred to me that shy, mousy me had the power to hurt Margo."

Christine wonders if the etiquette slip would have happened if she had had a socially savvier mom who could have warned her not to exclude anyone, for any reason, or who could have advised her on how to smooth over the situation. "I wonder why I never wrote Margo a note or invited her to come over to the pool another day to apologize about the party mix-up, to make sure she knew that it was an oversight. I definitely paid for it the next year, in middle school. Margo was immediately part of the most popular clique, while I hung out with a different and less socially powerful group of girls. She had set her sights on me as her enemy and took every opportunity she could to be really mean to me. The bus in particular was painful because I had no friends on our route, and Margo did. She'd whisper things about me to them and taunt me. I was miserable."

Margo was clearly a "mean girl." "But when I look back," Christine says, "I think of the pain I could have avoided if I'd had better skill in navigating the social world or a mom or dad who could have helped me. I think my parents were apt to say nothing or to tell me that Margo seemed like a spoiled, bratty kid that I wouldn't want to be friends with anyway. In one sense, they might have been right about Margo, but clearly they were also missing the realities and nuances of a shy girl's social world."

Without Darkness, There Is No Light

Children's friendships can be a refuge from a troubled home or a troubling clique. Yet some of friendship's power to shape us and teach us is within its less sweet traits. For starters, "friendships aren't always symmetrical," Rubin says. "Some say that parent/child relationships are vertical and friendships are horizontal. This just isn't true. Many friendships have a dominant person and a less powerful person." Kids are always grappling with hierarchies—a good preparation for real life. A boy who is controlled by big brothers at home might get a boost from lording it over buddies outside his living room, and one who rules his parents' roost might be the runt of *his* class, complete with a bossy friend to put him in his place.

Olsen, of the University of Memphis, has explored a curious social structure within the dark side of friendship: one-way pals. These asymmetrical relationships used to be written off as "noise" in the data, Olsen says, but once he started counting them, he was shocked by their prevalence. When he asked several hundred kids in grades three through six to list their friends, about 12 percent of the time one person put down another as a buddy, while the other independently reported that he disliked the nominator. (In 10 percent of cases, the dislike was mutual.)

"Children who are believed to be in friendships they aren't aware of are very high on social functioning," Olsen says. You might think these kids are two-faced or engaged in some sort of power struggle in which they are trying to manipulate other children into thinking they are their friends. But Olsen finds no malice behind the incongruency. "They're nice and they make everyone think they are a friend. And it turns out that making everyone think they are your friend is very adaptive. You benefit from treating the world well." Those on the other side of the asymmetry—the kids who think they are in a friendship when they aren't—tend to have deficits, such as an inability to pick up on social cues.

The rival adds a tart flavor to friendship. She is often respected and liked, even if her skills in a particular area are coveted. A rival can push a child to work harder and excel and is thus generally useful, if vexing. But if her rival is always on top, it can be hard for a kid to flourish. One of my friends, an amazing jazz and pop singer, had the great misfortune of sharing her small hometown with an ever-so-slightly better singer her age. Year after year, the two girls faced off in local competitions, and year after year, my friend would lose, until the losses became a part of her self-image and she developed stage fright. It wasn't until she escaped to a bigger city (and entered a prestigious music academy) that she realized she could develop her unique talents and thrive in a diverse artistic landscape—way outside her rival's shadow.

Another character who taints the romanticized picture of childhood as an innocent oasis of playtime and laughter is the enemy. Is he the opposite of a friend? Not quite, says nemesis expert Maurissa Abecassis, of Colby-Sawyer College. But an enemy does provide opportunities for personal and social growth, just as friends do. Abecassis has found that up to 75 percent of people have had an enemy by the time they are young adults. Some of these "mutual antipathies" are simply two kids disliking each other but not taking up much space in each other's minds. Other enemies become a more menacing repository for a kid's conflicts. "One way to cultivate an emerging sense of self is to draw contrasts with others," Abecassis explains. "You could dislike in others a characteristic you have in yourself that you can't accept, or a characteristic you're jealous of that you don't have."

A more agonizing variety of enemy is the former friend (meaning a "breakup" of some sort, not merely a drifting apart, occurred). Two people who trust each other and have disclosed much to each other are exponentially more vulnerable than two acquaintances, says Abecassis, if the relationship goes south, say, from jealousies or other negative feelings that they can't surmount. Once the first betrayal strikes, mutual hate can ramp up to levels as strong as the love between the friends once was. If the two former pals are enmeshed in the same greater peer network, the falling-out will be even more devastating, as mutual friends may be forced to take sides.

Though it's common to have at least one enemy at some point, children who seem to collect antipathies, Abecassis says, are obviously contributing something to these dynamics above and beyond the common forces that repel some children from each other. And though enemies help kids learn how to handle conflict, Abecassis stresses that one doesn't need an enemy to learn these lessons. "But one does need a friend," she says. "Without a friend *or* an enemy, there is no potential for disagreement."

Having an enemy is not necessarily scarring. Rejected children, if you recall our typology, are vulnerable to making enemies out of whole classes and as such have tipped the scales in previous studies testing for the effects of antagonistic relationships. Once you remove those who have experienced this broader form of peer rejection from the equation, you can see that the long-term negative effects of garden-variety enemies are very slight.

Which brings us to the king of all antagonistic childhood figures: the dreaded bully. A bully is not a friend at all, of course, though a bully and his victim are involved in a relationship—one characterized by dominance and submission. One recent survey of third graders in Massachusetts concluded that a full 47 percent had been bullied at least once, while 52 percent said they'd been called a mean name or had been teased in a hurtful way, and 51 percent said they had been excluded or ignored by pals within the past few months. Alarming as these numbers seem at first glance, the very survey design betrays a tendency to associate normal peer behavior (teasing or occasional ignoring) with full-on bullying. An ongoing bullying relationship is obviously detrimental and requires adult intervention. Learning to deal with slights, power struggles, and strong personalities, however, is just part of life and should happen largely without adult interference.

By the same token, gossiping and teasing are used to bond with friends as often as they are employed to fuel conflicts. Such behavior starts early. Preschoolers regularly exclude playmates from activities and use threats to get what they want, as in "If you don't play the daddy, you can't be in our group." Adults, while more subtle and sophisticated, continue to engage in relational aggression. "Our moms tell us not to gossip," says Grace. "But they do it with their friends all the time! It's hard not to do it."

It's difficult even for the most conscientious child to hold her tongue, because we really are built to gossip. It's the human equivalent of primates' spending hours grooming each other, and it really does encourage the ever-important group cohesiveness. It's one of those many instances where instinct dances uncomfortably with socialization: Few parents can refrain from gossip, yet they must tell their children not to engage in it—in order to nudge the child's natural inclination toward the most civil use of gossip possible. In other words, kids who learn how to gossip just enough to get in with others, but not so much that they are dubbed cruel or inflict real damage on the targets of gossip, will do well socially. Telling them not to gossip at all is the only way to rein them in to a moderate level of dishing that is probably optimal in the adult world. Kids don't always get subtlety; as such, who can blame moms for preaching "gossip is bad" rather than "gossip is natural and useful, but you must continually monitor yourself when flinging dirt about others to avoid going too far"?

Contrary to the stereotype that boys settle such matters with their fists, it's not just girls and women who use silver tongues to wield social power. A recent study shows that while it's true that boys are more likely than girls to become physically violent with other kids, boys are just as likely as girls to also use psychological and verbal tactics on peers.

Technology has also changed relational aggression (a topic I'll tackle in more detail in Chapter 7). "Home used to be a refuge from bullies and tormentors," says Abecassis. "Now they are ever present. And the social media sites are like an elaborate game of telephone: They allow one jab or insult to take on a life of its own, under the watch of the entire peer network."

When asked about "mean girls" at her school, Ella recalls just one nasty incident. A girl named Jessica had argued with her best friend, who didn't want anything to do with Jessica anymore. Rather than just break off the friendship, the other girl shored up allies and began taunting Jessica in front of everyone, calling her a loser. "It was hard to see Jessica sad and alone," Ella says. "But it seemed like I was the only one who felt bad. So I went over to her. Once I did, some other girls did, too, and now we're all friends."

Ella's story bears witness to the protective power that just one friend has for a picked-on child. Kids are less likely to be victimized in the first place if they have just one friend, and those who have been taunted for a year are not at risk for developing difficulties such as anxiety or a tendency to lash out if they have a best friend. Friendless victims do show more of these difficulties after a year.

Even more run-of-the-mill social upsets are mitigated by just one good pal. A Canadian study of cortisol levels in fifth and sixth graders' saliva (a biological indicator of stress) found that both boys and girls who were in the presence of their best friends had lower levels of the hormone after stressful social interactions (such as being teased on the playground) than those who were by themselves.

A KID COMES WITH A TEMPERAMENT as well as parents who hold some sway over whether or not he'll make and keep friends. The links he forms with pals will always be in the context of a bigger group into which he must fit or suffer the consequences. He'll likely experience taunting from some people and will weather slings and arrows from even his closest companions. Whether his social world is a relatively happy bubble of thrills and challenges or a threatening, ongoing nightmare, it's not just a passing situation but rather a prologue to adulthood.

What kids themselves might be heartened to know is that the lessons of social maneuvering are rarely mastered completely. Even the grown-up who has an engaging and easygoing manner, who emerged from the battles of the playground successfully, and who had the benefit of close friends with whom to build up empathy and resolve differences will still be ruffled, at age forty or fifty, or beyond, by the aggressive rival at the office and will still deeply feel the stings of slights by—and the muck of misunderstandings with—her friends. Social life is unavoidable (even most recluses have to confront this fact eventually), and some pain comes included with its many joys.

CHAPTER 4 *Friendship in Adolescence: Confidantes and Partners in Crime*

IT WAS THE '80S AND LYDIA WAS twelve when the beautiful, brazen Rachel moved to her school in London, Ontario. Rachel was a dancer. She acted. And she was from Toronto—the big city. "Anyone from Toronto was impressive to begin with," Lydia says, "and I was just fascinated by her." Rachel achieved instant popularity, intimidating others with her quick tongue. "She could tear people down in a second," Lydia says. Lydia had her own group of nerdy friends but flew far beneath Rachel's social radar. Her friend crush developed from a distance.

The smartest kid in school, Lydia had long lived to please teachers and amass gold stars. This rage to achieve was in part a ploy to bring peace and happiness to a home where her mother had essentially left and her father, a professor, fumbled through single parenthood. The problem, Lydia realized that summer between eighth and ninth grade, was that her shtick wasn't working. Stellar report cards weren't bringing her parents closer to each other or to her, so she completely altered her modus operandi: "I decided I was going to care about being popular and having fun." Her new number one aspiration: Befriend Rachel.

When a skater boy developed a crush on Lydia (much to her surprise), Rachel, who was interested in the guy herself, finally took notice. "That was the game-changer," Lydia says. "If boys started paying attention to me, then Rachel and her friends had to pay attention to me."

Lydia was in uncharted social territory. "Ninth grade was the most stressful year of my life because at every moment I was trying to negotiate being in the right place, talking to the right person, saying the right thing," Lydia says. "Rachel was a mad monarch in a way and I made her laugh—I was her jester." It was all so strategic, yet Lydia never felt she was being inauthentic. "In trying to fit in with Rachel I was finding out who I was." When the two began to hang out one-on-one, Lydia fluttered with an excitement that surpassed anything she'd felt for a boy. She had indeed accomplished her mission, and she and Rachel became closer and closer throughout high school. Never mind her father's disapproval.

"My father hated Rachel with a passion. He thought she was a terrible influence on me. I was really insulted by this because he didn't know anything about her. I guess it's ironic, because in a way I was incredibly influenced by her, but as far as smoking or drinking, I was going to do those things anyway. He didn't understand why I was so taken with her." Rachel *did* dress provocatively, flaunting her "developed" figure. She wasn't bookish, though she was smart, and she didn't care at all about school. College was never a goal for her. And she talked back to her parents constantly.

But as for Rachel's real influence, she became a behavioral model for Lydia to study. "She had this great laugh that I emulated. I asked myself, 'Why do all of the guys like Rachel?' I broke it down. It was her mannerisms and the way she blew them off." Rachel implanted a fundamental idea into Lydia's head: that some people were interesting and some weren't, but that you could learn to *act* interesting. Rachel also sharpened Lydia's already keen sense of humor. "I developed more of an understanding of irony and learned to process moments for their humor, both to entertain people and to earn her respect." It wasn't that she imitated Rachel outright; that would have invited scorn. It was more that watching Rachel gave Lydia permission to be less constrained and more commanding of others' attention.

When Lydia announced she would delay going to college for a year to travel out west with Rachel, her father stopped speaking to her for several months. Rachel's indulgent parents took over, driving the girls to the airport and stuffing their pockets with cash.

Once the two eighteen-year-olds settled into a hostel in Vancouver, their youthful dream morphed into a more fraught reality. "I just remember this feeling that Rachel was more unstable than I imagined," Lydia says. "And she was finding me more uptight. I was the buzz kill because I didn't want to go to a seventh bar at five a.m. There was a limit for me. There was no limit for her. And more and more she needed me to take care of her."

One day Lydia overheard Rachel complaining to her mother on the phone about Lydia's being a stick-in-the-mud. "She made me sound seventy years old," Lydia remembers. She confronted Rachel, sparking the biggest fight the two had ever had. "I told her, 'I'm just different from you. I'm not doing anything wrong, and I'm not antisocial. I'm just not you.'

What I learned was that confronting her was something I could handle. In my family, it seemed like if you did that, the walls would fall down."

After struggling to hold on to jobs for months, the friends decided to take a cross-country train trip. When they hit Winnipeg, all their belongings were stolen: their passports, train tickets, Lydia's glasses. It was a sad ending to a heady time. They made their way back to Ontario, and Lydia went off to college that fall. Rachel returned to Vancouver and fell into a heavy drug scene. Soon she had tattoos on her face and was working as a stripper. "Our lives completely diverged," Lydia says. "She was hard to stay in touch with, but when we did talk, we still felt close, as though we were going through similar things."

The tale of Lydia and Rachel encapsulates the strong attachments and dramatic push and pull that characterize teenage friendship: Lydia found in Rachel a charismatic personality to admire, but in doing so she discovered reasons to appreciate the core of her own character. Rachel was an emotionally expressive person who showed Lydia a way of being that was different from what Lydia had known inside her family's house. She even, for a time, supplanted Lydia's family of origin. In a sense, Lydia's father was right. Rachel did lead Lydia temporarily off one path and onto another. But he underestimated his daughter's ability to make her own choices in the end. Lydia's father was also right about the path he suspected Rachel would end up taking, which is always a disappointing outcome for a teenager who longs to prove her parents wrong. For Lydia, the friendship was the perfect arena for wrestling with the key issues of adolescence: boundaries, identity, how to be with the opposite sex, how to break free from one's parents. If friendfluence were divided solely by age, its biggest chunk would likely be allocated to the teen years.

Leaving Childish Ties Behind

Good playmate potential is an important criterion in kiddie friendships. Preteens start to share deep and secret thoughts with their pals, and in adolescence, intimacy and emotional support are friendship's essential components. Teenagers want friends who can help them become who they want to be. As an adult, you still need to feel that your friends reflect your identity (or your desired identity), but that drive was probably more urgent when you were an adolescent.

In fact, to the average thirteen-year-old, friends are just as emotionally supportive as parents, and to seventeen-year-olds, they are more so. A corroborating piece of evidence for the importance of teenage peers over parents comes from a Canadian researcher who asked college students about their lives thus far. One question she put to them was, "What above all else made you unhappy?" While 9 percent recounted a sob story about their parents, 37 percent wrote about damaging experiences with peers.

As children enter middle school, they begin to talk to their friends about the more difficult aspects of their family lives. "Children under the age of ten are extraordinarily loyal to their parents," write the authors of *Best Friends, Worst Enemies*. By age thirteen, they begin seeing their parents' flaws and share astute observations about them with buddies. Long psychological conversations with pals are even more frequent among high schoolers—especially girls.

A key difference between middle school and high school

emerges as late adolescents form romantic attachments, which sometimes take precedence over friendships. Still, it's all a continuum: The skills kids use to keep up their same-sex friendships are further developed through their romantic ties. And whereas middle school friendships are often situated in tight cliques, in high school, teens disperse and tend to float among different groups. All-girl and all-guy gangs from middle school merge into mixed-sex friendship clusters. Finally, while small cliques might break up between middle school and high school, larger "crowds" are easily identifiable in most high schools. As Kenneth Rubin describes it, the high school crowd is "a reputation-based collective of similarly stereotyped individuals. Kids are assigned to one crowd or another by the consensus of the grade or school population; though friendship-based subgroups may form within a crowd, its members aren't necessarily good friends and may not even spend much time together."

DO YOU THINK YOU DIDN'T FALL far from the tree, so to speak, in how you've turned out in life? Independent researcher Judith Rich Harris makes a counterintuitive argument: Except for the genetic material they provide, parents exert little influence on their own children; rather, children are socialized and shaped by their peer groups. She doesn't think individual friends have an influence, either, though she acknowledges that friends are probably part of that peer group. Harris's main point—that parental behaviors don't determine kids' fates—is worth entertaining in our era in which overprotective "helicopter" moms and dads believe they can push their teens in a particular direction by sheer force of will and resources.

If you had a rough upbringing, you might remember your friends (and their homes) as safe havens. For adolescents who have antagonistic relationships with their parents, friends are indeed a valuable buffer. Such kids with supportive best friends don't fall prey to negative self-perceptions as do those who are socially isolated and suffer from bleak family situations. Great friends can also teach social and relationship skills to teens who are mistreated at home.

James Olsen says that while it's hard to separate out the influence of friends from the influence of parents (and other factors), it's useful to conceptualize a quadrant: In the lower left corner you have teens with a bad home life and a bad peer life. These poor souls have no respite from disconnection and rejection, and the two circumstances probably feed on each other over the years. In the bottom right-hand quadrant you find kids with a negative family relationship but good friends, who provide a compensating effect on mental health. These teens might be independent types who have figured out how not to be enmeshed in a bad domestic dynamic, or perhaps they occupy a healthier social role at school than at the kitchen table. In the upper left-hand corner are kids with a warm living room hearth but fraught peer relationships. "Maybe it's not the kind of home life that values the things that the peers in a kid's town value," says Olsen. "Maybe the family is religious or not into sports or the status objects that one needs to fit in." In the upper right-hand quadrant are those lucky teens with supportive, functional families *and* strong ties to friends.

One wouldn't want to travel through adolescence alone, notes teen expert and clinical psychologist Carl Pickhardt, Ph.D. "The reality is that if you have one or a few [friends], that's all you're going to need." During my first year of high school in North Carolina I was really close to only one girl.

We got along well with others and were not ostracized despite being slightly nerdy; we just weren't part of any of the bigger groups. We had a blast eating lunch together every day and hanging out on the weekends, when we'd scour the mall for cute boys from other schools (I don't think we ever actually talked to any) and watched *Gone with the Wind* and other films that were surely out of step with our peers' tastes, but not horribly weird. I don't think either of us ever felt deprived or outcast by the small size of our clique.

By adolescence, teens have settled into a friendship pattern based on three variables, Pickhardt says. "Is the kid shy or outgoing? Is he insecure or confident? Is he solitary or social?" These traits aren't always linked as you'd imagine at first glance. Some outgoing and confident teens still prefer to be alone rather than the life of the party, for instance. Problems can stem from a mismatch that keeps kids from getting their needs fulfilled: The shy, insecure teen who wants to be social but can't manage it is going to be much more unhappy than the shy teen who only needs a few friends to feel connected to and accepted by his peers. Shy, insecure, and solitary teens have to do a lot more work to make friends, but they can do it.

Just having those few close reciprocal friends makes a teen more altruistic, sensitive to feelings, independent, and less likely to be anxious, depressed, or aggressive. And if an adolescent starts getting rejected by her peers, a close best friend acts as a cushion, preventing her from perceiving her social standing as being on a precipitous decline. (Kids who were in the "rejected-aggressive" category discussed in the last chapter might be more likely to be full-blown delinquents by the time they are adolescents. While Pickhardt is hopeful that any teen can turn around his social situation, the cumulative effects of being in an undesirable social category since childhood can be hard for teens to shake off.)

Developmental psychologists seem unanimous in their belief that good friends matter much more than popularity. A common teenage misconception is the thought that if only one were popular, he or she would have a lot of friends. "But when you're popular, you realize that you have a lot of people who want to be friends with you *because* you're popular," Pickhardt says. "Conversely, that is why being one of the 'uncool' kids is like a social disease. There's the sense that it's catching by association."

Adults don't tend to fear that simply being nice to someone who is struggling or in some way undesirable will in any way reflect upon them—except positively. But for teens with their shaky sense of self, there seems to be a perception of permeability between them and those they are kind to. It takes a mature and strong adolescent to walk over and sit with the morbidly obese loner in the corner, one who doesn't think her stock will plummet as soon as her tray hits the lunch table. When one kid is able to befriend an unpopular kid without suffering a social status drop herself, it makes it much easier for others to rush in and do the same. Consider this heartbreakingly familiar phenomenon that crops up frequently in teen movies and the like: Two friends are completely content to be together outside of school and are a perfect pal match. But as soon as they enter school and the collective glare of the peer group, the slightly higher-status friend, under the cruel spell that makes social status more important than individual attachment and happiness, will pull away from her dear comrade.

Children grow into teenagers just as they always have; the basics of developmental psychology are apparent across time and place. Adolescence also unfolds under specific cultural expectations and in reaction to social and technological change. American adolescents are especially peer oriented, compared to teens from other countries, for example. In the United States and Europe, teens spend as much as a third of their time with friends. But as Daniel Hruschka points out, kids in East Asian countries tend to have less discretionary time to pal around. In a study of eleventh graders, American teens spent 18.4 nonclassroom hours per week with friends, while Japanese teens spent 12 hours and Taiwanese spent just 8.8.

The Internet and the widespread use of computers and mobile devices have left some adults feeling that teenagers are entirely different creatures from previous humans—digital natives whose coded interactions with one another can't be cracked. Half of American teenagers are sending sixty or more texts a day, after all. Chapter 7 will sort out how friendship is affected by our tech revolution, but my basic conclusion on where teens land after this quantum leap is that social media sites and the ability to text or otherwise easily stay in touch do not fundamentally alter the way they interact with their peers; they merely amplify it. Thirty-five percent of teens still talk to friends face-to-face every day outside of school, even as they also communicate with friends via myriad electronic devices. But the teenage tendency toward rash decisions does mix dangerously with Web sites that can quickly and widely disseminate gossip and venom. Social aggression can strike at any hour of the day, not just when school is in session, and small slights and dramas can scale up to soap operas with the touch of a button. On the other side of the coin, if a kid with social deficiencies has only online friends with whom he can be himself, that's infinitely better than having no friends at all. In the era of social media and cell phones, the good elements of social life can be even better—and enjoyed by more who were previously excluded—but the bad elements can be even worse.

Who's Friends with Whom?

Aside from their being requisite to youthful surviving and thriving, teenage friendships have the distinction of being the genesis of three of the most successful rock bands in history: U2 (Larry Mullen Jr., age fourteen, put up a notice at school for musicians interested in joining the Larry Mullen Band); the Beatles (Paul McCartney, fifteen, convinced his friend John Lennon, sixteen, to team up with George Harrison, a fourteen-year-old who played in a rival band at the Liverpool Institute High School for Boys); and the Rolling Stones (Keith Richards and Mick Jagger met at Wentworth Primary School, where Mick was a serious student and Keith was a truant, and later reconnected as teens). More recently famous bands formed by adolescent pals include Live, the Wu-Tang Clan, Maroon 5, Red Hot Chili Peppers, the Beastie Boys, Fishbone, and Linkin Park.

Each of these bands could be said to sum up a certain attitude and aesthetic; by finding fellow musicians who agreed with that profile, the band members got not only identity support but an artistic vehicle for expressing and further aligning themselves with that attitude and aesthetic. Apart from enjoyment of music, teens seem to need to rally around bands, whether famous or based in a nearby garage, to see their style and outlook reflected and to find a crowd of friends and potential friends who are also drawn to performers who broadcast "This is who we are!" Adults might pick music purely for the mood and sound of the songs themselves; for teens, becoming a fan of a band is often tantamount to signing up for membership in a particular social group.

A Dutch study confirms that adolescents look for pals with personality styles and musical preferences like their own. The twist this researcher uncovered, though, is that young people gravitate toward others on the basis of their perception of a potential friend, not his or her actual character. Teens project images onto others, making friendly attraction a product of their own egos, as well as a matching game. These projections can be very accurate, however, even if they are based on nonverbal cues, such as attire. Angela Bahns, Ph.D., of Wellesley College, found that pairs of people can pick up on each other's shared prejudices quickly, based on cues such as style of dress. Multiple facial piercings, for example, do not best friends make, but they obviously communicate something more to potential friends than "I like to voluntarily put holes in my perfectly nice face."

Jesse Rude, Ph.D., and Daniel Herda tracked high schoolers for a year and found that compared to those friendships that did not stay intact, friendships that lasted were between kids who were similar in terms of GPA, school attitudes, and alcohol and cigarette use. Surprisingly, there was widespread instability among so-called BFFs: In this sample, about three-fourths of dyads "broke up" after a year. The authors concluded that the finding underscores the need for more research on the factors that keep friendship stable, as well as the factors that bring two people together in the first place.

Rude and Herda were most interested in the stability of interracial friendships. Prior research found that racial differences are more of a block to friendship formation than are class gaps: Students of the same race whose mothers have different levels of education are more likely to be friends than an African American and a white student. Yet when friendships between kids of different races form against these odds, they've shown to improve racial attitudes and to grant minorities more access to job and school opportunities. Since interracial friendships could be a powerful vehicle for social change, Herda and Rude reason, it's important to learn how they play out among teens. In their sample, though, interracial friendships were less stable over a year's time than same-race friendships, even when similarities in other realms were taken into account.

Meanwhile, Pickhardt sees advantages across the board for teens who bridge the gender gap. "I think cross-sex friendships are more common in small private schools, where social life is less scripted than it is in big high schools," he says. The benefits of cross-gender friendships are undeniable. "What kids learn from their own sex is very stereotypical information about the other sex. It's dehumanizing." As a consequence, Pickhardt believes, those who glean direct insights into the opposite sex are poised to have better dating experiences in the future.

While adolescents generally form tight bonds, the differences across cultures can be illuminating. Alexys, a confident and outgoing seventeen-year-old from Port-au-Prince, Haiti, often spends time with relatives in the United States and has noticed a national difference with regard to opposite-gender friends. "Here in the States if you go out with a guy, it's like you're dating," she says. "With us, it's more mixed. When I go out to a party or something, a guy friend always picks me up. And I like to hear what my guy friends say about other girls, like about what bothers them."

Alexys senses that friends are more intimate in Haiti, too, in part because social life revolves around families getting together with other families. She has known many of her friends as far back as she can remember, and her parents are close to most of her friends' parents. All those overlapping connections thicken friendship bonds. "We have parties, we go out for pizza or ice cream, but mostly we hang out in each other's houses because there is not much else to do. There are no movie theaters or malls in Haiti," Alexys says. "This makes us closer because we talk more."

The brutal earthquake of January 2010 that devastated Haiti scrambled Alexys's social world. Many classmates moved away. Luckily, she didn't lose anyone close to her, though several of her cousins were badly injured. Because her school was too damaged to open, she, her brother, and her mom moved to Miami for six months. (Alexys's circle is blessed with more resources than the average Haitian.) Her best friend, Chloe, with whom she had always been close—the two girls were born two days apart and had had all of their birthday parties together ever since—went along. "When we were living together in Florida, it was a good experience for both of us. Then, when we got back home, she started hiding stuff from me. She started saying things to people about me instead of telling me to my face. She was even dating someone, but I heard about it from someone else; she didn't tell me."

That same interconnectedness that forged a strong link to Chloe made the sudden shift in the friendship, which Alexys never could really account for or explain, all the more jarring. Though their friendships have a record of instability, teens might still harbor the child's fantasy (like Ella's in the last chapter) that their friends will be "forever." Knowing that such longevity is a rarity could help them deal with these painful rifts and ruptures. "When we first started growing apart," says Alexys of her former best friend, Chloe, "I couldn't sleep and I couldn't stop crying. We still have to see each other since our parents are best friends, but she's just changed. We used to go to every party together. Now I'll call her to see if she's going to one, but she'll say no." Now entering her senior year back in Port-au-Prince, Alexys luckily has other friends to rely on. "Since the earthquake, I've realized which people are really there when I need them."

Friends, Schools, and Neighborhoods

Having friends who care about grades can boost a kid's GPA. For instance, new middle schoolers with buddies who respect classroom rules do better on their in-class assignments than those with disrespectful friends. Those who latch on to troublemaker pals see slips in their grades within a year. Surprisingly, though, in a study conducted by the University of Oregon Child and Family Center, girls who were not doing well academically in sixth grade and then chose high-achieving friends ended up doing even worse by the end of the year. Perhaps, the authors reasoned, comparisons to their friends further eroded their confidence. Girls already doing well who chose high achievers as friends, however, did even better over time. They could handle the challenge of a grade-conscious peer group and in fact thrived within it. One of the authors, Thomas Dishion, Ph.D., emphasizes that the transition to middle school is pivotal. In another study, he examined the peer relationships of thirteen-, fifteen-, and seventeen-year-old boys and then looked at how well adjusted they were (defined by their level of work and school engagement and whether or not they had been arrested) at age twenty-four. Influences at age thirteen proved most predictive of adult functioning.

Judith Rich Harris, the researcher who believes parents don't exert much influence on their kids, cites a very important exception: Parents choose where their children live and go to school, and that, to a large extent, will determine the peer group that will influence them so much. "By living in one neighborhood rather than another, parents can raise or lower the chances that their children will commit crimes, drop out of school, use drugs, or get pregnant," Harris writes.

Consider a sample of so-called high-risk families of African American kids without fathers whose mothers bring in low incomes. In a study Harris recounts, kids fitting this profile who lived in poor neighborhoods were more aggressive than kids who lived in middle-class neighborhoods. But those with the exact same profile who nonetheless lived in middle-class neighborhoods were no more aggressive than the average child. Harris concludes that these teens adopted the norms of their peers, no matter what was going on inside their homes.

Moving might be a "cure" of sorts for kids in a peer group that doesn't value the same things mom and dad revere. But moving, especially if done frequently, is generally hard on teens. Making new friends and getting in with the right group take time under the best of circumstances, but for teenagers, a move often involves an additional task of parsing a fresh subculture with new rules and subtle signifiers that would mean nothing in another town or city. Each locale requires them to adapt to a new set of attitudes and ascend a new social order, perhaps from the bottom. "Kids who have been moved around a lot—whether or not they have a father—are more likely to be rejected by their peers; they have more behavioral problems and more academic problems than those who have stayed put,"

Harris writes. Because divorced or abandoned moms often take a financial hit, which results in their moving to a lower-income area, moving itself, Harris says, rather than fatherlessness, could explain why children of single parents sometimes have worse outcomes than those from intact homes. It's not that dad isn't around; it's that his income isn't, and that lands the kids in a rougher neighborhood with rougher friends to influence them.

Andy, seventeen, is a senior at a school in a posh New York suburb. The child of a single mom, Andy lives in an apartment that is perfectly nice by most standards, but among the smaller homes in his mansion-packed town. By choosing to be a smaller fish in a bigger pond, his mother has given him a huge advantage in our competitive society. His public school is filled with kids who are very serious about achievement. Perhaps because it's not natural to acknowledge the very air one breathes, Andy doesn't attribute his own successes (he's a straight-A student, a member of the track team, and has already been accepted to Cornell) to the influence of his peers. He figures the fact that both of his parents and his older brother went to college is what makes working toward that goal natural for him. The family standard can't hurt, to be sure, but neither can the fact that at his school, "it's all about getting into the best college." In fact, every single kid in his class will head off to a campus, "unless they take a gap year to travel abroad," Andy says. As smart and hardworking as he is, and granted that his parents and sibling set a good example, being surrounded by kids who view as a given the steps one needs to take to get into a great school makes it much easier for Andy to focus his natural drive and abilities.

In contrast, poor neighborhoods have high rates of teen crime and violence. And most adolescents commit crimes in groups. In fact, having deviant friends may be the best predictor of whether someone will engage in more serious forms of delinquent behavior in the future, making one's choice of neighborhood crew all the more important. One research team looked at the influence of parenting behaviors and deviant friends on delinquency among adolescents living in disadvantaged areas in Philadelphia and Phoenix. Friends yet again trumped parents: Hanging out with troublemakers predicted criminal offenses, while parenting behaviors—healthy or unhealthy—did not. Also, while you'd think that "social cohesion" would be a good quality in a neighborhood, in these impoverished areas, this factor made for tighter peer groups and even more delinquency.

When he was a doctoral student at the University of Illinois, Greg Dimitriadis, Ph.D., now a professor of educational leadership and policy at the State University of New York at Buffalo, worked at a community center in the heart of a poor and crime-ridden area outside Chicago. There he got to know two fifteen-year-olds, best friends he calls Rufus and Tony. Dimitriadis ended up tracking their lives for five years and wrote a book, *Friendship, Cliques, and Gangs: Young Black Men Coming of Age in Urban America*, that cuts through common tropes about kids growing up in rough places. "There are two stereotypes," Dimitriadis says, "*Hoop Dreams* and *Menace II Society*." The book underscores the tragic reality that a "good" outcome for young men in this context (essentially avoiding jail and death) is quite different from a "good" outcome according to social scientists studying teens.

Dimitriadis describes how Rufus and Tony saw themselves as part of a neighborhood "clique"—their word—that was separate from the gang dominating the area. Tony was in the gang for three years, during which time the clique was a respite from the stress of the gang, with its violence and the paranoia it brought on as he feared revenge at the hands of rivals. Far from pressuring his friend to join him, Tony was proud that Rufus did not get sucked into the gang. Rufus, who was the sole caretaker of his sickly mother, relied on Tony for practical help and relief from his many responsibilities.

Gang involvement is often characterized as serving a kid's needs for belonging and love. Really, Dimitriadis says, "self-directed kids are the ones who get into gangs. It's self-interest that motivates them; they are seeking protection, respect, girls, or money. Gang allegiances are tenuous, and membership, except at high levels, is not that formal. The construct of loyalty can fall apart pretty quickly when problems erupt."

Tony's experience is a case in point: Rufus and the clique, not the gang, were his sources of fraternal support. But gang involvement, for Tony, was a means of making it in the community at large. For his part, Rufus was sometimes made uncomfortable by the attention adults gave him for resisting a life of crime, for being cast as the "good" kid. Dimitriadis concluded that the community center itself played an important role in the friendship of Rufus and Tony. "Schools stratify kids; this community center did not," he says. "At the center, Tony was valued for his ability to handle younger kids. So it was a space for his best self to come out and where friendship was allowed to grow."

Teens are often wary of grown-ups, and most adults are annoyed with, intimidated by, or indifferent to adolescents. We are starkly segregated by age in our culture. As such, the friendship I found particularly interesting in this tale was the one between Greg, the academic in his late twenties, and Rufus and Tony, the urban teenagers. How did this alliance—crossing age, race, and class—bloom? "The connection evolved from the mutual demands we put on each other," Dimitriadis says. "I asked to interview them for my thesis, and they started relying on me for day-to-day errands. For them, it was more about the material scarcity and daily challenges they faced. Depending on where someone is in their life, they might need moral support from a friend, for example. But if you have to go get groceries but don't have a car, that's the immediate need."

I wondered if he felt like a do-gooder when lending a hand to kids with such a heavy deck stacked against them or if he thought of himself as someone they could look up to, someone to plant in their heads visions of the world outside of their mean streets. "I would never say I was a role model. That would have been presumptuous," he says. "They were role models for me because I saw everything they had to go through, each day. And I didn't feel noble; it was a feeling of satisfaction that one feels when forming an authentic friendship." That Greg worked at the community center legitimized him in the eyes of the other adults in the neighborhood; otherwise, "there might have been concern about this older white guy hanging around," Dimitriadis says. All three are still in touch; Rufus and Tony are in their thirties now, with kids. While formal mentorship programs could conceivably have similar results, Dimitriadis doubts it. "Friendship has to be chosen on both sides. That's why it's difficult to assign mentors. It's a flawed model."

Around the same time but across the economic spectrum, at the elite, WASPy boarding school Groton, another cross-age friendship grew, this one between a fifteen-year-old and a fifty-nine-year-old chaplain. "Jack Smith was my Bible studies teacher," says Matt Hutson, now thirty-four, "and I was an outspoken atheist." Hutson listed all of the contradictions contained in Genesis on the second day of class, and Smith, an Episcopalian minister, loved his gumption.

"Matt is someone who is more intellectually able than me. Even back then," says Smith, now seventy-seven years old. "We would be talking about something simple, like going to class, and he would 'kick it up to the sky,' as we would say, and start getting onto serious subjects like global warming. It's intriguing to me, how he thinks."

"He became a friend instead of just a teacher," says Hutson, who selected Smith to be his academic advisor for the rest of his high school years. During that time, Hutson suffered from depression and was briefly hospitalized for it. "Senior year I was missing a lot of classes because I was sleeping in. An antidepressant I was taking was making it hard for me to get out of bed. Jack started coming to my dorm room to get me up before chapel."

"When I went in to wake him up," says Smith, "Matt would ask me to give him a word problem to make sure he was actually awake. This was funny for me because I hate math and started dreading coming up with word problems for him every morning! But we made it through the year. That's the sort of thing that cements a friendship."

"One Saturday morning I didn't feel like facing the world. So I went back to sleep after Jack woke me up," Hutson says. "Later when we spoke, he said, 'You're not an asshole, but that was an asshole move.' I thought, 'Yeah, you called it.' "

"Matt once said, to me, 'Who am I? I'm the kid whose best friend in my high school is the chaplain!' He was accepted by his peers and even appreciated by them. But he wasn't exactly one of the guys," Smith says.

Hutson remembers that he was self-critical, so he didn't

throw himself into the social mix as much as he would have liked to. "It's not that I was picked on," he says. "It's that I set myself apart. I wished I'd been more of an insider, but I didn't really know how to take social risks or how to include myself in interactions. Jack provided stability for me at Groton. He's very wise, and he really understood the teenage mind, so he knew how to deal with me in a flexible way."

The two now e-mail regularly, and Hutson has visited Smith, now retired, on Cape Cod a couple of times. They have dinner together when Smith comes to New York, where Hutson works as a science writer. "Being clergy, I sometimes feel that people are needy," says Smith. "But with a friendship like this, it's more balanced. And I always did get more from Matt anyway. It was never just me giving to him, even back then. I admire him so much. He has his depression, but he carries on wonderfully."

"Jack told me I was one of the bravest people he knew," says Hutson. "I still can't fully accept that compliment. For someone I respect so much to have that level of respect for me is a really special thing." Without Smith's support, including his wake-up service, Hutson isn't sure he would have even graduated from Groton, a counterfactual that likely would have changed the course of his life for the worse. Peers are so important to adolescents, but among their ranks is not the only place one can find a friend—and thus a measure of salvation (even for a nonbeliever).

The "How" of Teenage Friendfluence: Contagion, Pressure, and Training

We know that peer pressure peaks during the teen years. (Friends are more influential than the peer group generally, but scholars often use the blanket term "peers," which could also include classmates or just those in the same cohort.) What probably concerned your parents when you were an adolescent was whether or not your friends were going to get sweet, innocent you to do "bad" things. That's what most of the research focuses on, too. Many behaviors spread among groups of friends, not unlike contagious diseases. How does this happen? Does it unfold just as in those alarmist after-school specials, where a pushy kid forces a quiet one to take drugs, leading immediately to addiction and deviancy? Of course not. As we probably all remember from those high school drinking parties, the road to mischief is lined with much more subtle signs than "everyone is doing it!"

Take the provocative term "deviancy training," for example. This is one mechanism that explains friendfluence among teens, and it's essentially kids sitting around and swapping stories about "the time I got messed up" or about "how cool it would be to rob so-and-so's store." In other words, deviancy training refers not to a deliberate indoctrination but simply to a conversation that promotes deviant acts among its participants, who laugh, smile, or nod along to these stories. Thomas Dishion, from the University of Oregon, has been studying deviancy training for years and notes in one of his papers that the process is unconscious for the teens, who are motivated to hold an audience of friends and gain their companionship. The acts themselves that lend content to the conversation are seemingly secondary to those needs.

Dishion and colleagues even found that just measuring the length of a "deviant" conversation between pairs of friends accurately predicted whether they were "early-onset antisocial," "late-onset antisocial," or "successful." Those pals who gabbed about a norm-breaking theme and couldn't seem to switch gears to other conversation topics were likely to show more problem behavior throughout the rest of their teen years. If you catch a fourteen-year-old (that's the age where resistance to peer pressure is lowest) talking excitedly and repeatedly about high jinks, this line of research implies, don't be surprised if he's antisocial at age twenty-four. The more practice a kid has in the art of deviant talk, the more likely he'll need to keep gathering materials and stories to preserve his forte.

You might wonder if it's not the talk per se that "trains" friends to make trouble. Could those already predisposed enjoy the talk more and seek each other out? The leading teen development expert Laurence Steinberg, Ph.D., a professor of psychology at Temple University, believes that as in the nature/nurture debate, we're never going to tease apart the effects of selection from socialization in friendship, so it's better to accept that both forces are at work. "Parents should know that their kid is not an angel getting corrupted by a group. He has something to do with it," Steinberg says.

Steinberg and his team have recently revealed another view of how the peer pressure machine operates, by peeking into the teenage brain under the influence of friends. They wondered whether the tendency to take risks in the presence of friends (think: speed limit adherence when a teen is alone versus when his crew is in the backseat) has to do with the incentive processing, or reward system, in the brain and/or a cognitive control system, which enables impulse control and deliberate comparisons of choices. They measured brain activity in these systems while adolescent and adult participants played a simulated driving game (winners received a monetary prize, so the motivation to take calculated risks was built in). Subjects were asked to bring two same-age, same-sex friends to the experiment. Some played alone while others played knowing their pals were watching on a monitor in another room.

Adults didn't show any difference in the reward region of the brain whether their buddies were tracking their game or not, nor did they play the game differently, in terms of blowing yellow lights or in other ways taking chances in the hopes of getting ahead. But adolescents did show more activation of the reward center when their friends were watching. Correspondingly, they took fewer risks in the driving game when they were playing alone. (Girls were less interested in risks than were boys, but the increase in the risks they took when friends were watching equaled that of the boys.)

A compounding effect makes sense since "some research shows that activating the brain circuitry with one type of reward—say, drugs—will prime the circuitry to be sensitive to other rewards such as money," Steinberg says. The study, he adds, changed the way he looks at peer pressure. "It's not an explicit thing where kids are pressuring others. These kids weren't even in the same room. The reward of risk was simply stronger when friends were present."

Another form of talking that fuels friendfluence is co-rumination—sitting around and obsessing over a problem, as though spilling a high quantity of words about it would make it disappear. If you're a woman, I'm sure you remember doing this with your teenage friends (and you probably still do it to an extent). Men and boys are far less likely to fall into co-rumination, and that is protective for them; it can cause teenage girls to become depressed and anxious. Amanda Rose, Ph.D., an assistant professor of psychological sciences at the University of Missouri, Columbia, and an expert on co-rumination, points out that girls think talking about their problems will make them feel better. Sometimes it even does in the short term, but over time it bogs them down. The upside is that girls who co-ruminate do tend to be close to each other. If girls knew co-rumination was a downer (or if their parents could recognize it), they could be persuaded to go out and do something or meet up with other friends to interrupt the "Why did he ask Becky out and not me?" marathon.

When Rose surveyed boys about their feelings on opening up and talking about problems, she assumed they wanted to do so but felt too embarrassed to try. It turns out they simply do not endorse it as a smart strategy. Rose recommends finding a healthy middle ground: Male friends could learn to get a few suggestions or supportive words about dilemmas from one another, she says, while female friends could stand to focus more on solutions than on endlessly dissecting their worries.

Related to co-rumination is the state of depression itself, also found to be contagious among friends. Having a friend with depression makes it more likely that a teen will become depressed herself. Researchers speculate that what catches on, more specifically, is the depressed teen's attributional thinking style: "I got a bad grade because I'm terrible at math and nothing ever goes well for me" versus "I got a bad grade, but I'll study harder next time."

In addition, talking about body image and eating behaviors with friends can, over time, create a shared dysfunctional view on weight concerns and appearance. A friend's engagement in extreme weight-loss strategies predicted a girl's later use of the same tactic, no matter what her fitness level. Obesity also spreads through teen friend networks, again perhaps because of a subtle infiltration of values ("it's okay to eat fast food every day") that then get reinforced as more pals adopt the unhealthy behaviors.

You won't be shocked to learn that eighteen-year-olds assigned to live with roommates with similar heavy drinking histories went on to drink more alcohol together over the course of the year. But a recent social networking study uncovered a twist on peer pressure and drinking: While his or her own friends influenced how much a teen drank, a teen's girlfriend's group of pals, or boyfriend's band of buddies, are even more influential in setting the drinking standard. If your daughter starts seeing a guy whose friends binge drink, she is an incredible 81 percent more likely to binge drink herself. And if your daughter merely has a one-sided crush on such a guy, his friends' imbibing behaviors are even more influential on her; she has more reason to try to "get in good" with them by being just like them.

Peer contagion has the unfortunate effect of rendering some youth programs not just ineffective but iatrogenic, meaning that kids are worse off after completing the program. The seemingly benevolent-sounding Cambridge-Sommerville youth program, which sent high-risk teens to camps, had a counterproductive effect, for example. Deviancy training is easier, one supposes in hindsight, when all the deviant kids are thrown together in a cabin in the woods. Dishion also found that middle schoolers selected for a therapeutic psychological intervention in a group format showed increases in smoking and delinquent behavior for at least three years.

Of course, individuals do have the power to withstand peer pressure, and adults can help. Teens with a lot of self-control (basically, the ability to stop their impulses in the face of those potential rewards that lit up the brains of average teens in

Steinberg's driving simulation experiment) do resist some of these peer contagion effects. Grown-ups who keep an eye on kids and provide routines and rules also reduce their kids' susceptibility to negative peer influences.

Reforming Peer Pressure's Bad Rap

Adults are influenced by their friends all the time, but they rarely think about it, or refer to it, as "peer pressure." A person who is not susceptible to peer pressure at all is in for a very difficult and probably unfulfilling life. Group norms keep us together, connected, and ethical much more than they hold us down or suppress our morality.

Joseph Allen, Ph.D., a professor at the University of Virginia, has a refreshing view on how teens shape one another, and he has done a great longitudinal study of teenagers that illustrates his stance. "The problem in adolescence is not that these socializing influences exist, but that they are increasingly from adolescent peers whose values at times run counter to those of parents and other adults," he writes. "Even within adolescence, however, peers clearly can have positive socializing influences—teaching one another about everything from handling the give and take of group discussions to providing information about positive after school activities." Allen and his colleagues conclude that "strengthening adolescents' positive connections to both peers and adults may help us steer these influences in more positive directions."

Adolescents who are popular not just because they are perceived to be powerful but also because they are sincerely liked by many others are usually very well adjusted—from their mental states to their social skills. They also tend to be more influenced by the dominant values of their group. Some of those values may be ones that adults understandably disagree with—drinking to appear "older," for example. But others, like treating other kids less aggressively, get the grown-up seal of approval. So the most well-adjusted teens are the ones most likely to adopt the norms of their group—a hallmark of socialization. Their ability to take in what the group expects shows sensitivity, Allen argues. They might engage in a bit of rebelliousness, but they might also model, say, good hygiene and turn taking.

Given that teens are passing on values to one another, Allen proposes an intriguing remedy for the negative parts of this socialization process: Help teens adopt better values by allowing them to make meaningful contributions to society, alongside adults. Though teens can display wonderful mental and physical powers (in other eras, they commanded large armies and governments, Allen points out), they are separated out from adult society and left to do hypothetical tasks that don't impact others. Kids who participated in a community service program and small group discussions to help them process their new activities showed 50 percent reductions in school failure, school suspensions, and teen pregnancy rates over the course of the following year, compared to a control group. Friends who help others and begin to feel that their actions are important will start spreading better values through the efficient peer contagion machinery.

Diana Marino, a thirteen-year-old who lives in the Bronx, seems to be living out the many positive aspects of peer contagion. Though she's just now starting high school, she already volunteers at a local arts and education center, Casita Maria. Like the participants in Allen's study, Diana makes connections to older kids, adults, and her community at large by helping out at the center. An outspoken and well-liked kid, she fits the "nice" definition of popular, rather than the "feared/disliked" variety.

Last year her closest friend, Elena, started cutting herself regularly to ease her anxiety. "It was hard for me because she didn't want me to tell anyone," Diana says. "The guilt and stress were eating me alive. After a few months, it started getting worse, and she was always covering herself in a coat." Diana told an adult, and Elena, while upset at first, was later grateful to be sent to counseling. "It was hard because I felt like either way I was letting Elena down, whether I did tell or didn't tell," Diana says.

The two went to a rather tough middle school, where a lot of girls break into physical fights at recess. "I've never been in a fight, even though my friends are like 'Jesus Christ, Diana, why don't you fight?' I don't want to get in trouble or get a record. A lot of kids go to juvie, and I don't want that." Diana tutored Elena in social studies, to help make sure Elena would graduate from middle school. "I pushed her and pushed her. I told her, 'You have to do this!' " As she starts high school, where her older friends from Casita Maria advised her to be herself in order to make good friends, Diana is already thinking about college. "A lot of people say I'm too mature," she says, pointing out that she is perhaps going against the grain in her neighborhood. In another departure from her peers, she's not obsessed with Facebook. "I check it when I go home, but then I play guitar or read." Whether as she moves through high school Diana will continue to be a good friend who influences others positively while resisting negative influences herself is an open question, but to judge by her sparkling self-confidence, it's a safe bet.

Reading, Writing, Relating

"A good friendship in adolescence is golden. If you have one, you'll have the capacity to create other relationships in life, including romantic ones," says clinician Carl Pickhardt. "But we have a funny culture; we don't teach kids this stuff. We think they 'know' how to do friendship, but it's not always true." If he were to teach a course on friendship to teens, Pickhardt would spell out that good friends can be relied on, that you can confide in them, be authentic with them, and that you should share common interests and have fun with them. He would have teens ask themselves questions about a friendship in order to evaluate it: Do you like how you treat the other person? Do you like how he or she treats you? How do you treat yourself in this relationship? And do you like how he or she treats him- or herself in the relationship?

Parents wondering how to encourage their teen to make friends must make it clear that there is nothing fixed about him preventing him from doing so, that it all comes down to behaviors one can practice, Pickhardt says. "A kid in this situation should use fear as an advisor but do the opposite. The fear is telling him not to look anyone in the eye, not to speak to anyone. It's a good formula for not making friends. Start doing the scary behaviors, and things will get better."

TRYING ON NEW BEHAVIORS, WHETHER consciously or unconsciously, is something that teens really excel at. You might remember getting lured into a crowd that wasn't quite "you," or you might recall being a ringleader who set the fashions and agendas for your friends. Either way, you probably have some theories as to how those adolescent experiments in acting and being shaped your life. Our minds often wander back to those intense and eye-opening teenage friendships—even if we no longer speak to the friends who forged our tastes or habits or worldviews. Adolescence is for feeling and experiencing many things for the first time—as such teen life is more vivid, more exhilarating, and more devastating—and the friends who accompanied us during those Technicolor moments are often burned into our memories and our very being.

CHAPTER 5

The Incredible Perks of Friendship

FOR THE PAST FORTY YEARS, Richard Levinson has spoken on the phone with his friend and fellow trial lawyer John every single day. They've been in touch at least several times a week since living together in the army in 1957. And since about 1972, they've bet on every football, golf, and basketball game on TV. "This year we bet a total of a hundred thousand dollars. We figure one of us will come out on top, by about fifty dollars. John jokes that the only winner in this enterprise has been the phone company," says Levinson, who is now eighty and still works at the New Jersey firm Levinson Axelrod, where he is a partner.

Richard and John discuss books, politics, and their legal cases. "We're different," Richard says. "I'm an effusive, emotional person, and John is a laid-back, cynical kind of guy. But deep down we're similar in that we both feel for people and for humanity in general, and I think despite our argumentative, joking banter, that's a united force."

John isn't Richard's only extremely long-term close pal. In fact, you might call Richard a "super friend." There's the Nobel

Prize–winning doctor whom Richard clicked with in the sixth grade and meets for lunch each spring. There's the uproariously funny friend who unfortunately passed away fifteen years ago; Richard and his wife still have dinner with that friend's widow about once a week to recall his antics. And there's his childhood buddy with whom he meets every year on December 7, to commemorate the fateful day when the two were taken by their fathers to a Giants game in 1941. As the boys, then ages nine and eleven, sat in the stands, announcements were made with increasing urgency for armed forces personnel in attendance to gather outside the stadium.

How has Richard managed to maintain all these special ties for so long? He modestly points out that friendship is two-way, and he's lucky to have friends who care to make the effort—and manage to avoid major conflicts—as much as he does. He also pleads sentimentality: He loves to remember the past, and these long-standing pals are the ones who can indulge his nostalgia. Finally, Richard didn't marry his wife, Susan, until he was forty-nine. All those years of being single, during an era when most people tied the knot in their early twenties, made these relationships especially important to him.

And for the past thirty-one years, Richard believes all of these rich friendships have enhanced his marriage. "We double-date and go out with other people two or three times a week," he says. "It's nice to see Susan react to other people and have others react to her, and both of us just enjoy it a great deal. We're extremely outgoing, and meeting with friends contributes to our happiness." Richard even argues with Susan over who, of the two of them, is closest to *her* old friends.

Richard and Susan have passed on the value of long-term friends to their son, who is now thirty. "He's very tight with his college group. He's also gotten a lot out of our friends over the years. Many of them are interesting personalities or have achieved great things," Richard says. Sustained exposure to different adult role models in his parents' home surely benefited the younger Levinson.

"To me, friendship is everything," says Richard. Though he has blessedly escaped any major work or health crises in his life, he's been given a sense of security by his solid friendships, knowing others would be there for him. He has also had the satisfaction of helping them with their troubles and guiding their children. Perhaps most notably, he has enjoyed hours of comfortable conversation with his pals. "Talking with these people is like getting into a pair of old slippers. I don't have to worry what they think of me or what I think of them. No one is trying to prove anything. Friendships have been the most pleasant aspects of my life."

As Virginia Woolf put it, "Some people go to priests; others to poetry; I to my friends." Friends can be our main source of moral support, midwives to our dreams, and generous suppliers of love, humor, and understanding. The man who gives a kidney to a dying buddy, or the woman who acts as a surrogate mother to a friend who's unable to bear children: Such dramatic acts of friendship make for excellent newspaper stories. But more often, the positive influences of friendship are less headline-worthy but still powerful: the daily phone chat and sports bet, the clarifying e-mail from a friend articulating her reservations about your new boyfriend, or the small unsolicited favor that reminds you during a bleak moment that you are not alone in this world. These gestures are the ones we notice and remember, but meanwhile such friendships constantly affect us in all sorts of ways we've probably never realized—physically and intellectually, as well as emotionally.

Sharpen Your Mind

You might imagine that befriending a brainiac could make you smarter over time, but it also turns out that talking to any friend for a few minutes improves your cognitive function. Researchers at the University of Michigan found that a brief chat upped subjects' scores on tests of memory and the ability to suppress distractions. Interestingly, talks that were competitive in nature did not show those benefits, so don't hitch a ride with your debate team rival on the way to the LSATs.

The researchers noted that their finding shows neural overlap between social functioning and general intelligence. This is a profound point because it supports the idea that friend skills are tantamount to general success skills and that IQ is not as separate from "EQ" as we might have thought.

Creative ability is much harder to pin down than IQ, working memory, or other mental attributes, but many scholars argue that anyone can learn to be creative—that is, can train him- or herself to come up with novel solutions to dilemmas. Friends can help jump-start and sustain this form of thinking as well. While we romanticize the idea of the lone genius, friendship often spurs creativity in the arts and sciences. Thirty years ago Michael Farrell, Ph.D., professor of sociology at the State University of New York at Buffalo, started collecting cases of what he calls "collaborative circles"—friendship groups that spawned original ideas and projects. "Creativity is linking two ideas that have not been linked before," Farrell says. "I'm interested in the kind of creativity that comes out of social interaction. It's not coming from one mind; it comes from people sharing their ideas."

Farrell's idea to examine how friends incite intellectual

breakthroughs came from his own experiences studying group therapy. Farrell saw group therapy sessions, which can be very effective, as ways for friends to give each other feedback and insights that dismantled their defenses. As a result, they felt a more authentic connection to whatever work they were doing.

In his book *Collaborative Circles: Friendship Dynamics and Creative Work*, Farrell features a group of French Impressionists—Sisley, Renoir, Monet, and Bazille—that exemplifies how friendship circles become hotbeds of artistic output. Bazille, Farrell argues, was the gatekeeper, who saw something in each of these young painters. Monet became their charismatic leader who talked them into painting outdoors, a big step for the studio dwellers. Soon they began to form a common vision of their work, positioning themselves outside the artistic establishment.

Farrell identified specific interactions between friends that preceded creative accomplishments. A case in point occurred in August 1869, when Renoir and Monet sat alongside a river and painted the scene side by side. It was at that moment that art historians believe the two discovered impressionism. "Renoir and Monet were looking over each other's shoulders," says Farrell, "and were engaged in what I call instrumental intimacy, the ability to share one's wildest, craziest ideas and bounce them off the other person in ways that generate new ideas that wouldn't have come to either of them if he were alone."

Instrumental intimacy can be hard to achieve among colleagues and coworkers who don't have the trust and goodwill that friends develop. Farrell points to academic departments where colleagues unfortunately "fear humiliation for sharing a stupid idea or feel paranoid about people stealing their unformed yet good ideas."

Modernist giants Matisse and Picasso were two other artists whose friendship influenced and improved their work. The Matisse and Picasso exhibit that debuted at the Tate in London in 2002 and was shown at the Museum of Modern Art in New York explored their creative dynamic from the time they met in 1906 to Matisse's death in 1954. "When one of us dies," Picasso wrote to Matisse, "there will be some things that the other will never be able to talk of with anyone else."

"For artists in general, and Picasso and Matisse in particular, exchange is vitally important," says Elizabeth Cowling, an honorary fellow at the University of Edinburgh and the author of *Interpreting Matisse Picasso*. "Each came to regard the other as the single most important artist. Each had a strong sense of his own genius and the fact that the other matched him." Picasso even said once that "Matisse is the only one whose opinion matters." And, Cowling says, "They were equally hurt if one said something disparaging about the other."

Friends choose friends like them, as we know, and Picasso and Matisse found their equal in talent and ambition. They are thought to have had different personality styles, but Cowling came to the conclusion that they had much in common. "We think of Picasso as impulsive and violent and Matisse as sober, restrained, and reasonable," she says. "But Matisse was tormented. He had a violent temper, sleeping problems, and was every bit as dramatic as Picasso. I think they recognized in each other an urgency and driven nature."

They were keenly aware of their different strengths. Picasso saw that Matisse was a "genius with color," and Matisse would have said that Picasso was a brilliant draftsman, Cowling says. Matisse sometimes acknowledged Picasso's sheer inventiveness but likely agreed with others that it sometimes shaded into gimmick. Cowling notes that because they picked themes and styles that spoke to the other, Matisse came out looking more uncontrolled in the exhibit and Picasso, more controlled. It's a nice metaphor for many friendships: People come together because they are like-minded and then grow even more like-minded, taking on each other's qualities over time.

The mutual influence started early on, in 1912, when Matisse began experimenting with Picasso's signature cubism. "The greatest paintings he ever did were the ones where he was battling with cubism," Cowling says. This in turn caused Picasso to push cubism even further. After World War II, Matisse started using paper cutouts. After Matisse's death, in the '60s, Picasso began creating cutout sheet metal sculptures, clearly in response to Matisse's cutouts. "I can't say whether the influence was conscious or unconscious," says Cowling. "I've had conversations with artists where I thought a particular influence was blindingly obvious and they say they weren't thinking about it at all. How influence works is mysterious." Picasso also did studio paintings, knowing the studio was a subject associated with Matisse. "He used a monochrome palette, which was actually a touching way of honoring his friend."

SUCH COLLABORATIVE FRIENDSHIPS ARE NOT unique to artists. In 2002, Nicholas Christakis, a professor at Harvard University who is a physician and sociologist, began working with James Fowler, a political scientist who is a professor at the University of California, San Diego. The two used their respective areas of expertise to explore social networks and published the best-selling *Connected: The Surprising Power of Our Social Networks and How They Shape Our Lives,* based largely on their analysis of the Framingham Heart Study. The two mapped how 5,124 residents of Framingham, Massachusetts, were linked from 1971 to 2003 and added a fresh view on how friends, neighbors, and spouses influence one another's behaviors, emotions, and outcomes in life.

Not only do the two study friends, but they are friends. "It's been the most rewarding professional collaboration of my life," Fowler says. "I do more of the number crunching and Nicholas does more of the writing. But when I bring him a statistical report, he instantly gets it. And I get it when a turn of phrase needs to be changed. So we can let the other shine, but have enough understanding of one another's strength to truly help each other."

Be Happier

It's no news flash that friends make us happy, but Meliksah Demir, Ph.D., a professor at Northern Arizona University, has drilled down to reveal exactly what about friendship warms our hearts. It turns out that companionship—simply doing things together—is the component of friendship that most makes us happy. And the reason friends make us happy, Demir has concluded, is that they make us feel that we matter.

Demir also found that compared to subjects in fifteen other countries studied, Americans rated their friendships as highest in quality. It could be that our emphasis on independence is really independence from family, but not from buddies. Relatedly, the more we live apart from parents and as singles or divorcées (more than half of households are made up of these constituents now), the more we need friends to fulfill many roles for us, and consistently so.

Nobel Prize winner Daniel Kahneman, Ph.D., of Princeton,

and colleagues conducted an innovative study about a decade ago that captured people's happiness "in the moment" as they went about their daily lives. They found, controversially, that time with friends is even more enjoyable than moments with spouses or children. When asked directly, most people who have them do consider their spouses, and certainly their children, to be the most important people in their lives, but that doesn't mean they are happy doing laundry, working to pay bills, or changing diapers. Kahneman's work doesn't suggest one should ditch one's family responsibilities to hang with friends all day long—that would no more lead to long-term happiness than eating doughnuts all day in place of substantial nutrition—but it highlights the importance of spending time with friends as a respite from life's relentless duties and pressures. Kids bring an overall sense of meaning and purpose to many people; friends are refreshing oases that give them the energy to cope with the hard business of raising a family.

Kahneman also found that commutes to work are common happiness detractors. Perhaps they'd be much easier to endure with a friend in tow, since boring and arduous tasks are better completed with pals. Strong relationships, including those with friends, are in fact the best predictor of one's general happiness level.

It makes perfect sense that having someone you care about, who cares about you, and with whom you share enjoyable talks and activities would boost happiness. But another mechanism by which friends spread joy is emotional contagion. Humans evolved to pick up on one another's feelings quickly, and contagion might have been a safety mechanism: If someone in your tribe looks scared, you get scared yourself and run accordingly.

If someone looks happy, you get happy and bond with him or her over the positive feelings; after all, your life depends on those bonds.

In their network analyses, Fowler and Christakis found that you are about 15 percent more likely to be happy if one of your friends is happy (overall, not in any particular moment). Even if a friend of your friend is happy, you're 10 percent more likely to be in a contented state. "We found that each happy friend a person has increases that person's probability of being happy by about 9 percent. Each unhappy friend decreases it by 7 percent," they write. Since these stats imply that happiness is more contagious than unhappiness, they conclude that "the more, the merrier" holds true, despite what is usually said about quality over quantity in friendships.

They also found that an additional friend amounts to two fewer days of feeling lonely each year. "Since on average (in our data) people feel lonely forty-eight days per year, having a couple of extra friends makes you about 10 percent less lonely than other people. Interestingly, the number of family members has no effect at all." (Many who dislike the holiday season would agree that one can feel very lonely in the company of relatives.) Even spouses don't relieve loneliness as much as friends do, according to Fowler and Christakis's calculations. And siblings don't affect one another's loneliness levels at all.

You've probably heard that religious people are happier than nonreligious people. With all due respect to the joy that a personal relationship with the Big Guy in the sky can yield, some social scientists believe that the aspect of religious involvement that makes the devout more satisfied with life is (drumroll) friends! "Friendships built in religious congregations are the secret ingredient in religion that makes people happy," conclude researchers Chaeyoon Lim, Ph.D., and Robert Putnam, Ph.D.

Close friendships predict church attendance and even the strength of one's belief in God. In his book *Vital Friends: The People You Can't Afford to Live Without*, Tom Rath theorizes that churches and other faith-based organizations, along with any club, sports team, or civic group, are "in the business of cultivating friendships, whether they realize it or not." We might consciously cite a purer soul, a healthier body, or a love of a hobby as our motivation for joining in, but the underlying motivation and reward might just be the friends we meet and that crucial sense of belonging they provide.

Know Yourself Better

"I had a very bad childhood," says Charlotte Cook, who grew up in California in the '60s, the daughter of a Polish mother and a German father, both of whom were Holocaust survivors. "My parents never overcame the scars of their experiences," she says. Her mother, who was abandoned during the war, was mentally ill, and Charlotte suffered abuse in their home, which was set on a dead-end street with no other children. "I came to college not having any skills as far as making friends was concerned."

But at Berkeley, during her sophomore year, she met Terry Looper. "She civilized me," says Charlotte, who was shy and also unnecessarily suspicious of others. "She kept popping me into social environments and nudging me to join in. She even dragged me to her parents' house, where a family actually sat around and talked. This was an amazing phenomenon to me. She had standards and wanted me to do well. She had a profound effect on me."

Terry was an art student but eventually became a social worker. She was warm, stylish, charmingly dramatic, and delighted in any creative endeavor. She helped teach Charlotte to become more trusting and social, not via direct lectures on how to behave but simply through time spent together. "Terry showed me how to make a friend and be a friend, to grow through the scary stuff and celebrate the unexpected," Cook says. "If I got frustrated with someone in my world, Terry would talk through the whole incident that bothered me."

When she met Terry, Charlotte was in a promiscuous phase, dabbling in drugs—par for the course in that era, but self-destructive all the same. Again, Terry didn't preach, but employed a gentle Socratic method to challenge Charlotte. "She forced me to at least be honest with my thoughts." Once, when Charlotte was having a typical young adult identity crisis, wondering who she really was and what she truly believed, Terry said something like "Think about all the things that you like, that you care about. All of those things live within you just perfectly. *You* are the integrating factor."

Charlotte began to realize that she didn't have to be one thing or conform to a single type. "I remember a flow of relief washed through me when she said that," Charlotte recalls. Terry's words encouraged her to continue to embrace the different parts of herself and to follow her tastes where they led. "Today I live in a classic California Craftsman house filled with art from Japan, Korea, China, and Vietnam, as well as art from India and the Southwest, with '60s rock posters and Swedish and Italian modern touches. And in my head I enjoy the same integration of thoughts and philosophies."

The two kept in close touch over the years until Terry died of lung cancer in 1991, when she was in her early forties. Charlotte continues to miss the friend whose life was cut short. "She was like a greenhouse for my personality," Charlotte says of Terry, whose spirit is detectable in everything from Charlotte's furniture to her values and ideas.

Friends like Terry can help us sort out who we really are and encourage us to become new and improved. Those we meet as young adults help us to forge our identities, overcome painful pasts, or even just traverse the typically rocky transition of becoming a grown-up. As many have noted, the time between adolescence and adulthood has been steadily expanding, thanks to an economy that keeps people from taking on grown-up responsibilities and a trend away from marrying young, thus spawning a whole new age category: emerging adults. Friend networks are largest during this time and get smaller as people settle into careers and love lives. In this period between studenthood and the solidity of adulthood, friends can be especially important. Psychologists Carolyn McNamara Barry, Ph.D., and Stephanie D. Madsen, Ph.D., write that during this time, "Friendships provide feelings of worth as well as opportunities for story-telling and frank discussions about religion, life aspirations, moral dilemmas, and relationships."

Emerging adults report that they feel closer to their friends than to their siblings. One reason may be that they don't feel a need to differentiate themselves from their friends as they often do from brothers and sisters. (There's an evolutionary explanation for the phenomenon of sibling rivalry: Throughout human history, siblings have had to compete for resources and attention from Mom and Dad. Even in well-off, modern houses where each kid gets his own bedroom, entertainment system, and college fund, the tendency to find a way to shine in comparison to one's siblings is strong and sometimes brutal. No wonder young adults, with all their insecurities and anxieties, prefer to open up to people outside of their own families.)

For adults of any age, friends can provide an existentially rewarding gift: They can truly know us, sometimes better than we know ourselves. Specifically, friends are better at describing our behavioral traits than we are, says Simine Vazire, Ph.D., a psychologist who runs the Personality and Self-Knowledge Lab at Washington University in St. Louis. "Friends can assess whether we are funny, dominant, or charming better than we can," she says. They may not be better than we are at knowing what we are feeling and thinking, unsurprisingly, but they are superior at guessing our IQs. (Incidentally, it's often the case that we judge ourselves as *less* intelligent than we are.)

The reason friends know our behavioral traits and IQs better than we do could be simply that we don't see ourselves clearly from the inside and/or that these self-judgments can be threatening to our self-esteem, Vazire says. It turns out that those who are more aware of their behaviors, however, are better liked. Here's where friends can yet again elevate our lives: They can increase our self-awareness (and, presumably, our popularity level).

My friend Adelle and I often discuss this precious aspect of friendship, which is also a minefield of sorts. A sensitive, empathetic person will take pains to protect her friend's feelings from being hurt, and that involves the telling of white lies. (In fact, one study found that close friends tell fewer lies per social interaction than others, but that they still do lie quite a bit. The catch is that lies among friends are more likely to be altruistic than self-serving ones—the kind we're more likely to tell acquaintances and strangers.) As we analyze the foibles of others, Adelle and I often say, "But you'd tell me if I were doing something like that, right?" Would we? I think we *are* honest with each other—it's a major strength of our very close relationship—but Adelle is so sweet and supportive, and so capable of seeing the complexity of a person's motives and actions rather than judging them as good or bad, that it's hard for me to imagine her exposing any delusions I might have about the ugly sides of myself, even though she knows I value that kind of feedback—in theory, at least.

I remember walking briskly ahead of my two orchestra mates once in high school, for the painfully clichéd reason that I was embarrassed to be seen leaving with them in the busy parking lot. Violin was a big part of my life, and my fellow violinists were smart, funny, wonderful people. But not terribly cool. In that moment, they gave me feedback in the best way possible: They laughed right at me and said, "Look, Carlin is embarrassed to be seen with us!" Their stark honesty and, at the same time, their sense of humor and lack of anger, proving that they were indeed above and beyond self-consciousness about any pecking order, shook me right out of my snobbish behavior and made me realize I was lucky to have them.

So how can you get better feedback on your behavior and personality from a friend? "Soliciting the information directly won't work," says Vazire. "Instead of asking, say something about yourself and then watch for your friend's reaction. I once said to a friend, 'Most people think I'm too serious.' My friend just sat there in response, confirming that it's true. Some people might deny these things out of politeness, but you will probably be able to tell if they are just being polite or vociferously denying your suspicion about yourself." Making an effort to know your friends better (as well as leveraging their insights to learn about yourself) will make your relationship less frustrating and less conflict ridden. It might be that even in close friendships, the "honesty" that takes place is really a coded form of communication, and the ability to send and receive such messages separates "super friends" like Richard Levinson from other people.

Inspire and Get Inspired

Friendship is often the special catalyst required for us to reach our aims. Whether you're trying to finish college or run a marathon, having a friend beside you, pursuing the same objective, eases and quickens the journey. One study provides a particularly touching metaphor for the concept while proving its strength: Researchers had subjects wear backpacks while standing at the bottom of a hill. They then estimated the hill's steepness. Some were next to friends, and others stood alone. Being with friends made subjects perceive the hill as less steep; being with long-term friends decreased their estimate of the hill's steepness even more.

Tom Rath writes about a study he conducted for Gallup in the early '90s on homelessness: "It was clear that alcoholism or a dependency on methamphetamines was more of a symptom than a root cause. In most cases, the relationship with a bottle or needle was precipitated by the collapse of a close relationship with a friend or loved one. The men and women who remained homeless for decades had something in common: a lack of healthy friendships. They were more 'friendshipless' than anything else—being without a home was just the most obvious and visible part of their plight." Rath goes on to tell of one woman who escaped her situation after befriending a woman who "expects me to be somebody."

Surrounding yourself with friends who hold you to a high standard and expect a lot from you might just be the best way to move ahead in life, whether you're starting from rock bottom, like the people Rath studied, or you want to achieve something spectacular and take your talents and abilities as far as they can go.

Just having friends nearby can push you toward productivity. "There's a concept in ADHD treatment called the 'body double,'" says David Nowell, Ph.D., a clinical neuropsychologist from Worcester, Massachusetts. "Distractable people get more done when there is someone else there, even if he isn't coaching or assisting them." If you're facing a task that is dull or difficult, such as cleaning out your closets or pulling together your receipts for tax time, get a friend to be your body double. "She could just stir the martinis," says Nowell. "She doesn't actually need to help in order to be helpful."

One incisive comment from a friend can be enough to inspire us, allow us to see our situation in a different light, or even drag us out of a particularly dark place. Ian Anderson, a British man who lives in Norway, worked as an aid worker in rural Uganda in the late '90s. "I would sometimes get overwhelmed by the apparent futility and desperation of it all," Ian recalls. One time a crowd of people came to Ian's house in the village, begging him to help a sick relative. "I looked at the old lady and saw that she was skin and bones, and put her in the car," he writes. "Half way to the hospital there was a commotion in the back seat as she died."

Ian, whose project built maternity health units, feared that the tragedy would reinforce local biases against Western medicine. Moreover, conditions were so bad in the village that someone died nearly every day. Late one night, though, Mark Allies, who had joined the project and become Ian's friend, told Ian that in order to keep sane, he looked at small things really closely. "When was the last time that you *really* examined a leaf, a spider's web, a strawberry, or a bird's nest?" Mark asked Ian.

"As a grown up," Ian says, "I had stopped noticing familiar things and had failed to appreciate the beauty that surrounds us. Looking for the perfection that does exist kept me going, especially in light of all the suffering I was witnessing. Mark taught me that there actually is order in the world, if you look for it."

As much as friends alleviate suffering, we are often willing to suffer for them, perhaps even more than we would endure for our own sake. Freya Harrison, Ph.D., of Oxford University, asked subjects to do squats against a wall for a small payment. One time the money would go to them, and other times the money would go to four different people, including acquaintances and friends. When a close friend was set to win the prize money, subjects did the grueling exercise for about 1.5 times longer, on average, than they did when they stood to gain. Notably, in an earlier study Harrison conducted, people were unwilling to squat longer for extended family members than for themselves, putting friends in a special category as objects of devotion.

Suffering is a relative term here. Examples of the sorts of efforts people make for a friend might be standing in line for hours for tickets to see her favorite band or making ornate decorations for a party in her honor—something you simply wouldn't bother doing for yourself or even for a relative. (Maybe because relatives are expected to love us unconditionally, we feel a need to "prove" our love to friends, who can get away more easily with ditching or downgrading us.)

Any psychologist will tell you that the best way to alleviate sadness and get positive feelings flowing is to do for others. Friends let us give to them, help them, listen to them, look out for their interests, and even sacrifice our time, energy, and resources. This is yet another gift they offer: the chance to continually care for them and reap the rewards of extending ourselves.

Live Long and Stay Healthy

The evidence of the salutary effects of friendship is so strong and is linked so clearly to common killers like heart disease, cancer, and obesity that one of the smartest health care policies never discussed on the Senate floor could seriously be an initiative to encourage and nurture friendships. And as anyone who has picked up a healthy habit only after teaming up with a pal can attest to, obligation and commitment to a friend can be far stronger than our own willpower.

Consider the obesity crisis. Most Americans are overweight or obese now, and losing that weight is a preoccupation produced by both vanity and fear of death. In their network analysis, Christakis and Fowler found that if a person's friend becomes obese, it nearly triples her risk of becoming obese herself. Even friends who live far away can influence our weight, meaning that it's not simply a shared environment (or a case of birds flocking together) that causes obesity to spread among groups of pals. It seems that an active relationship is necessary for one to "catch" obesity from a friend; if you don't consider someone a confidante, but she considers you one, she's much more influenced by you than vice versa.

Weight gain likely spreads through friends via a process of shifting norms. If we visit a friend who is ten pounds heavier, we might slightly alter our view of what kind of body size is acceptable. If our gang meets at a fast-food restaurant every Friday night, we'll begin to believe this is normal behavior, whereas another group in the same town might consider it taboo.

Of course, the influence can go the other way. Friends who get healthy set the stage for our own body makeovers. "A friend doesn't have to tell you he feels good because he's started jogging," says Fowler. "You'll notice it over time. And while you might not copy him directly and start running, you might start eating less instead. So the behaviors are different, but the outcome will be the same."

Fowler strongly suggests *not* dumping your chubby comrades, by the way. "Having a friend is more beneficial than not having one, no matter their potential influence on your weight. So instead, turn this principle on its head and be a good influence on your friend. Help her make positive choices and remember that your actions have a ripple effect on her friends, too."

The idea that behaviors spread via friendships has come under fire, because of the basic scientific problem of separating out correlation from causation and the effects of other people from the effects of one's environment. But other studies have corroborated the friend-weight link. For example, a new online experiment from MIT found that people were more likely to start using a "diet diary" if others using the diary in their online network were similar to them. That indicates that having a workout or weight-loss buddy "just like you" will motivate you to adopt better strategies.

Those recovering from mental illness can find succor in friendships with people struggling with the same issues. Jeff Bell, a San Francisco radio personality, for one, credits a friend he met sixteen years ago, Carole Johnson, with jump-starting his recovery from severe obsessive-compulsive disorder, an ailment she also had. "There is no substitute for friendships forged by common challenges," he writes in a tribute to Carole. "Carole and I have shared with each other our innermost fears, our triumphs and setbacks, and our heartfelt commitment to supporting each other through the good times and the bad."

Just spending time with your friends can reduce your stress levels. Specifically, women feel less anxious as the result of a progesterone surge that comes on when they feel close to a friend. Laughing with friends can increase physical pain thresholds by about 10 percent, and having a friend with you, or even merely imagining a friend, causes drops in blood pressure. Older adults concerned about memory loss should start calling up their friends: Elderly people with active social lives are much less likely to experience cognitive decline and dementia.

The ultimate argument for the positive influence of friends is their startling effect on our life spans. You may have read about a particularly striking study of breast cancer patients: Those who were socially isolated had a full 66 percent increased risk of dying compared to women with a supportive circle. Amazingly, having a spouse did not reduce the patients' chances of dying. The same held true for Swedish men followed for six years: Having a romantic attachment didn't decrease their risk of a heart attack, but friendship did. These results are strong counterpoints to the single person's stereotypical fear that she will die alone. Perhaps it's those without friends, and not the unmarried, who should worry more.

Julianne Holt-Lunstad, Ph.D., professor of psychology at Brigham Young University, did a meta-analysis of 148 studies and concluded that a lack of social support predicts all causes of death. People with a solid group of friends are 50 percent more likely to survive at any given time than those without one. Holt-Lunstad calculated that having few social ties is an equivalent mortality risk to smoking fifteen cigarettes a day and even riskier than being obese or not exercising!

Why is friendship such a strong match for the grim reaper? There is not just one answer, says Holt-Lunstad, but it likely has to do with how friends act as stress buffers, thereby ameliorating negative health effects, and how having friends might encourage healthy behaviors, such as eating well or going to the doctor when you're sick. "Friends could give people meaning and purpose in life, which could lead to better self-care and less risk taking," she says. "Throughout history we have relied on others for survival, protection, and resources. In modern times we have the notion that relationships are more for emotional satisfaction and have forgotten the physical piece."

Enhance Romance

Friends are often cast in opposition to lovers, but they turn out to fuel love more often than obstruct it. Friends, in fact, are particularly good at finding you someone new: They introduce people to 35 to 40 percent of their sexual partners. That said, young adults often ditch friends as soon as they take up with a new guy or girl, but Meliksah Demir warns that while a supportive romantic partner can have a greater effect on happiness for "emerging adults" than her parents or friends can, as soon as she's single, her friends are what determines her level of contentment. So keep your friends around, in case the flames of passion die down.

Besides, who better to discuss, analyze, and ponder your relationships with you than your friends? Unhealthy obsessive ruminators notwithstanding, friends are vital guides on the bumpy road to love. Because the heart wants what it wants, we sometimes ignore our friends' warnings, but we can usually rely on them to help put us back together when they're proven right.

Shelly, a thirty-year-old Canadian, was often told by her two closest friends and roommates that the guy she was seeing just wasn't good enough for her. "We would take walks most nights and rant about our lives so they knew all along what my problems were with him," she says. "It was not a physically violent relationship, but he was verbally and emotionally abusive. In retrospect I see that I had become a broken record and did not realize I deserved better." A few years after Shelly married her boyfriend, the abuse became physical, prompting her to move into a homeless shelter with her daughter. Her two friends convinced her to move back in with them until she could afford to find her own place. "Without them I believe I would have gone back to the situation I was in," Shelly says, "but they have kept me strong."

Established pairs are also well advised to cultivate and keep up their friendships, just as the happy couple Richard Levinson, the "super friend," and his wife, Susan, do. Richard Slatcher, Ph.D., an assistant professor of psychology at Wayne State University, induced couples to make friends with other couples in his lab. First of all, he found that those who stuck to small talk, perhaps unsurprisingly, did not become real friends, whereas those who talked about deeper or more personal issues were much more likely to get together for real double dates. Most intriguing was how couples rated their own relationships more positively after interacting with other pairs. Married partners fall into routine interactions and often fail to make the effort to entertain and please as they did when they were winning each other over. Putting your best self forward for new friends allows you to shine and to see your partner through new eyes as she shines, too. Maintaining older mutual friendships also strengthens the bond between long-term partners: Having people around who think of the two of you as a unit, who admire your relationship, and who expect you to stay together can sustain you through times of doubt or distance.

Advance Your Career

As sociologist Mark Granovetter discovered back in the '70s, the "weak ties" of a large friend network are critical to job seekers. The better you've been at introducing your friends to your other friends over the years, the broader and thicker your connections and the rosier your employment prospects. And once you start tasting success, you'll find your friends even more helpful. As Christakis and Fowler put it, "If you are rich, you can attract more friends, and if you have more friends, you can find more ways to become rich."

A little over a decade ago, sociologist Jan Yager, Ph.D., conducted a survey of friendships on the job and found that women had fewer friends and fewer friends in high places than did men in similar positions. She believes that women would go further if they made more casual office friends who could provide advice and perspective specifically on work issues, rather than try to make one or two close, personal friends. She also believes that women need to befriend those with more power, since we are judged, after all, by our work cliques.

Tom Rath, of the Gallup Organization, did extensive research on the role of friends at work and came to the conclusion that while most organizations don't encourage, or even actively discourage, friendships among coworkers, these ties are crucial to our well-being and performance. If you have a best friend at work, he writes in his book *Vital Friends*, you are "significantly more likely to engage your customers, get more done in less time, have fun on the job, have a safe workplace with fewer accidents, innovate and share new ideas, feel informed and know that your opinions count, and have the opportunity to focus on your strengths each day." Friends at work even make up for a less-than-ideal salary: People with close buddies are twice as likely as others to have a positive perception of their paychecks. If you can count at least three dear friends at the office, you are 96 percent more likely to be extremely satisfied with life in general. (Of course satisfied people might have an easier time making those friends, but still, making friends is a skill that can be improved with practice.)

In a finding that ties together the benevolent effects of friends on health and work, a twenty-year study concluded that the risk of death was significantly lower for people who reported a high level of peer support on the job. Befriend that guy one cubicle over, or DIE.

Solve Social Ills—or at Least Expand Your Social Horizon

We expect friends to help us. We probably don't anticipate helping the world by making friends. Yet friendship can be a powerful vehicle for social change, and this is the principle behind the many organizations that provide opportunities for Palestinian and Israeli children to become pals, or Irish Catholics and Irish

Protestants to form alliances. While friendships among people of different races are statistically rare in the United States, having such a friend lowers your levels of prejudice and even those of your other friends. Cross-race bonds also reduce stress and, more metaphysically, induce "self-expansion," allowing one to feel others' pains and joys more fully.

To answer the question "What is it like to cross friendship boundaries?" James Vela-McConnell, Ph.D., a professor of sociology at Augsburg College in Minneapolis, Minnesota, and the author of *Unlikely Friends: Bridging Ties and Diverse Friendships*, conducted interviews with people who had befriended those of other races, sexual orientations, and genders. People found these friendships to be eye-opening and developed more respect for people different from themselves. They claimed the difference was a source of bonding, not an impediment to it. Class, however, was not a category in the study because Vela-McConnell couldn't recruit enough subjects with friends from different social classes, a telling indication of what really constitutes a boundary in our society.

Dalton Conley, Ph.D., chronicler of race and class differences and dean of social sciences at New York University, agrees that friendship among those of different class backgrounds can be beneficial, if complicated. "I think there are two countervailing effects. If you're the lower 'socioeconomic status' person, having friends who are higher SES opens possible worlds that you didn't even realize existed," he says. "In my case, I went to Italy in high school with my friend's family. It was a huge experience. But it can also introduce relative deprivation where you didn't realize you were so badly off until you saw how the other half lives." Still, he advocates for assigned rather than chosen roommates in colleges, to increase a student's chances of making a connection across economic and racial divides.

Friends can even take credit for sparking large-scale social movements. Prior to Rosa Parks's act of defiance on the bus in Montgomery, Alabama, in 1955, others had been arrested for similar transgressions, yet they hadn't launched the civil rights movement as a result. Parks, it turns out, had many close friends who sprang into action upon hearing the news of her arrest, and furthermore, she had less close friends throughout the city, from different walks of life, enabling support for Parks to spread far and wide. Charles Duhigg, *New York Times* business reporter and author of *The Power of Habit: Why We Do What We Do in Life and Business*, says that a formula can be applied to many successful social movements: Friends are the first step; weaker ties, which build community and peer pressure, are the second; and a change in the habits of the movement itself, such as encouraging followers to take on leadership roles in order to keep up the momentum of the cause and sustain its growth, is the third. "A movement or even a company made up just of friends starts to fall apart eventually; you need those other steps," Duhigg says. "But those only become possible when you have the capacity to draw people into the organization, and the easiest way to draw them in is usually through friends."

In his survey of altruism and friendship, Daniel Hruschka concurs that while people regularly volunteer for noble causes, "such altruism often relies on recruitment through close relationships. And people generally favor causes when a close friend or family member is directly affected by the issue at hand."

THE NEXT TIME ONE OF YOUR FRIENDS invites you to join a crusade, you should do it (unless it feels cultish; we'll get to groupthink and other bad influences of friends in the next chapter). After all, you might just owe your good health, robust love life, sunny outlook, smashing career, and even your life itself to her.

CHAPTER 6

Bad Company: The Dark Side of Friendship

SHANE SHAPS STILL REMEMBERS the day her friend Claudia pushed her to climb the monkey bars when they were little girls in Louisville, Kentucky. Shane, who was terrified of heights, fell and broke her wrist. That type of bullying pressure from Claudia was not so atypical. Yet Claudia, who had been her companion since they were infants, continued to be an important—though, on balance, negative—part of Shane's life. "Our parents were close friends," says Shane, now forty and a social media consultant and the mother of two. "It was a codependent relationship. She bossed me around, and I let her do it. I'm not a complete victim." While in college, the two lived together for a couple of summers, and later they moved to the same building in Chicago after graduation. During that time, the four-foot-nine-inch Shane was thirty pounds overweight and struggling with low self-esteem and dating anxiety.

Claudia was fit and sporting a new nose, courtesy of a plastic surgeon. Shane would call her at work to talk, and Claudia would always say she was too busy. "Yet she was supposed to be my best friend," Shane says. Claudia did let Shane cry on her shoulder at times, and in fact Claudia seemed to need Shane close by in order to have the perverse pleasure of rejecting her or controlling her (and others) in tiny ways. When Claudia got engaged, for example, she asked Shane to be a bridesmaid and then sent the bridal party a four-page e-mail detailing the exact type of silver sandals each was to wear and other laughably specific requests.

A few years later, Shane got engaged to an on-again, off-again boyfriend and moved to New York, without a word of congratulations from Claudia. The next dip in the two friends' decades-long roller coaster ride was set off by an anniversary card: "I felt Claudia wasn't supportive of my engagement, but nonetheless, I sent her a card for her first anniversary. She called a few weeks later, furious that I hadn't sent her one. What was really happening was that she didn't like that my world no longer revolved around her. When the post office returned the card to me, I re-sent it and circled the original date stamp. She never called to apologize."

Continuing the cycle of dependency, Claudia threw Shane a shower in Chicago. The party was so awkward that Shane tried to demote Claudia from bridesmaid to guest. She responded by writing to Shane: "I'll be there and I'll be in the pictures. You can look at me for all time." Claudia's parents hosted a dinner party for Shane the night before the wedding. But once Claudia's father learned of the underlying drama, he refused to speak to Shane. The familial meshing that had kept the two women in the same web for so long had frayed, and Shane soon after had the courage to cut all threads between them once and for all. "Finally I said, 'To hell with this!'" says Shane. After thirty years of friendship, she initiated the "breakup," and the two haven't spoken in ten years. (Claudia is right there in the key nuptial photos, however.)

"I think our friendship never grew up, even as we did," Shane says. "We could not get past the fifth grade." Nonetheless, she says she learned from this destructive friendship. "I'm a stronger person in all of my relationships. I stand up for what I believe in. I have many friends, but I'm choosy. So despite all of the turmoil and tears, I thank Claudia for that."

Since friends are powerful influences in your life, they can just as easily have negative effects as positive ones, especially if they are not right for you, or if the dynamic between the two of you is unhealthy. Even loving, compatible friends can vex or hurt us from time to time, and the fluid nature of friendship sometimes makes its darker waters harder to navigate than conflicts in romantic or family relationships.

Christakis and Fowler have established that the overall effect of having friends outweighs any negative impact they might have. Yet in order to truly reap all of the friendly benefits outlined in the last chapter, you must take into account the ways in which friends can lead you astray. Some friendships, which are rarely purely bad but rather fall on a spectrum from "annoying in the way that all human relationships can be annoying" to, as the self-help gurus call them, "toxic," can exact emotional, physical, and spiritual tolls.

One case study from the art world conjures up a grisly scene compared with Renoir's and Monet's lakeside idylls. When painter Paul Gauguin met Vincent van Gogh, he was the more famous of the two. They shared an artistic vision distinct from that of the impressionists, and encouraged each other in their outsider views. Both produced masterpieces during the height of their friendship, even working together in a studio in the south of France.

But when Gauguin announced that he was returning to Paris, van Gogh crumbled; he even suffered a nervous breakdown around that time. Some claim van Gogh then stalked Gauguin with a razor and in a fit sliced off his own ear. Another recent account by two German art historians speculates that van Gogh didn't slice off his own ear after all; the injury might have been at the hands of his dear friend Gauguin during a sword fight. There's evidence that the two made a pact to keep the matter a secret.

We'll likely never know the truth, but the incident, however it actually played out, didn't completely, um, sever their bond. Van Gogh was forced to move into a mental hospital, but he stayed in touch with Gauguin. And from his self-imposed exile in Tahiti, Gauguin wrote letters to his friend, right up until van Gogh committed suicide. Even the most fruitful friendships can cause considerable pain, often precisely because the need and love the two have for each other are so strong.

Drift Enablers

I can safely guess that most of your friends will never contemplate de-earing you. Even if they are perfectly considerate and totally nonviolent companions, though, they still might have a subtly negative influence on you. You might not even be aware of it. I'm referring to friends who have goals, values, or habits that are just enough outside your ideals that they cause you to drift away from your core self and, consequently, from the aspirations most suited to you.

Gretchen Rubin, author of a blog and book called *The Hap-*

piness Project, defines the concept of drift as "the decision you make by not deciding, or a decision that unleashes consequences for which you don't take responsibility." If you're drifting, she writes, you might feel that you are living someone else's life and you might daydream a lot about escaping your circumstances. The key to this concept in terms of friendship is in Rubin's observation that drift often comes about when you do things because the people around you are doing them.

"We drift into certain decisions because other people approve of them. Your sense of what is right for you becomes clouded by what other people think is right. . . . You drift into marriage because all your friends are getting married. You drift into a job because someone offers you that job." Ultimately, this may bring on unhappiness. "Approval from the people we admire is sweet, but it's not enough to be the foundation of a happy life," Rubin writes.

Drift by friendship can happen when you grow and change while your friends contentedly stay where they are or when you haven't quite figured out your own talents and beliefs and are particularly susceptible to conforming to the values of those who surround you. One friend of mine regrets having spent his college days with a group of hedonistic frat guys and "second-tier jocks." Would he have been more thoughtful and open-minded as a young man if he'd had more conscientious and studious friends back then? he wonders. Of course there is not much to be done about a youth misspent with the wrong crowd (and something to be said for having had experiences with lots of different types of people). But the "right" group—not necessarily the most "popular" one, but a group that validates who you are and also projects an ideal version of yourself—can lift you up almost effortlessly over time. In contrast, staying with the wrong group, one that keeps you from a path to the fulfillment of your deeper needs and wants, will leave you walking in the wind, having to exert more and more effort just to move forward.

Avoiding drift shouldn't be confused with ladder climbing or using people to get where you want to go, as if the takeaway should be, "Dump your unsuccessful friends and get in with the 'it' crowd!" Treating people as mere instruments, while fine for networking, is anathema to making friends. Yet forging *sincere* friendships with people who bring out your potential will likely help you get ahead and bring you more contentment than could any amount of card swapping or backslapping. Of course there is no one right path we should be on, and there are multiple selves within each of us that we could develop. All we can do is to cultivate authenticity as much as possible amid circumstances we don't always control.

Stressors

The late Ray Pahl, a British sociologist, conducted a poll of about a thousand people and discovered that almost two-thirds identified friends as one of the biggest sources of stress in their lives. From the friend who criticizes your girlfriend to the one who clings to you with a needy grip, friendship is not as rosy as it is sometimes portrayed. In his excellent book *The Meaning of Friendship*, the journalist and philosopher Mark Vernon recounts more of Pahl's research: "Over a quarter say that friends are the main cause of arguments with partners and families; around 11 percent admit to taking a sick day in the previous year due to friendship problems; 25 percent say they can't cope with making new friends; and well over three quarters admit to wanting to lose at least 5 'flabby friends' as part of a New Year friendship detox—flabby friends being like those extra pounds that a healthy workout would shed." Running with the amusing concept of "flabby friends," I think it could be useful to think of those people as the ones who don't necessarily incite strong negative or positive emotions. Maybe you feel a vague obligation to keep up with them, but you don't feel nourished by their company and could perhaps invest your free time in more substantial friendships.

Friends who do stir up both affection and annoyance, however, are much harder to shed and to manage. Julianne Holt-Lunstad became interested in "ambivalent" friendships—ones that are a mixed bag of agreeable and disagreeable aspects. She had subjects wear blood pressure monitors all day for a period of time and record every interaction they had with others, and how they generally felt about those people. Unsurprisingly, encounters with people the subjects felt primarily positive toward were associated with their lowest blood pressure rates. Intriguingly, blood pressure was higher when they spent time with ambivalent friends than it was when they encountered people they described as primarily negative forces in their lives. "Because ambivalent friends are unpredictable, the subjects probably had a heightened level of vigilance while with them, which could explain the blood pressure spike," Holt-Lunstad says. "If someone is aversive to us, it's easier to discard what they say or blow them off." True "frenemies," as the tabloids would put it, might be less taxing than those sometimes great, sometimes not pals. Holt-Lunstad has also found that the more ambivalent relationships you have, the more at risk you are for depression and cardiovascular problems.

Why then, Holt-Lunstad wondered, do we keep these

ambivalent companions? Is it because, like Shane and Claudia, some are part of a dense friend or family network that we ourselves can't escape? That was the case with some of the subjects in Holt-Lunstad's follow-up study, but the top reason for keeping ambivalent friends around was not external but internal pressure. "They wanted to see themselves as the type of person who can keep friends," she says. "Or they already felt invested and didn't want to bail, or they held Judeo-Christian values such as 'It's best to turn the other cheek' that ran counter to dropping such a friend. Also, they justified these friendships by highlighting their positive aspects."

Subjects did distance themselves emotionally from ambivalent friends and believed they saw them less than mostly supportive friends. "But when we looked at the actual dates, they saw these ambivalent friends just as much." This study also confirms other scholars' observation that most people do not break up friendships outright—even if they are more negative than ambivalent in nature.

Teasing banter with friends is an entertaining release for many, yet outright negative and competitive encounters with friends might wreak havoc on your health by unleashing inflammation processes in the body. Jessica Chiang, a graduate student at UCLA, and colleagues had volunteers keep diaries documenting all of their good, bad, and competitive interactions (from games and sports to work or academic rivalry to interpersonal competition—e.g., vying for attention at a party). She then measured the subjects' levels of pro-inflammatory cytokines. Higher levels of these cytokines were found in those who had told more negative and competitive tales in their journals.

"Inflammation is a healthy response," Chiang says. "We

need it to heal wounds, for example. But activating that system when you don't need to, in absence of physical injury, is dangerous over time. Those who suffer from chronic inflammation can develop cardiovascular problems, arthritis, and depression." Chiang notes that leisurely competition, such as playing games or sports, did not increase inflammation, while the other forms of competition analyzed did. "The media coverage of toxic friends might be an exaggeration," she says, "but over years, an accumulation of social stressors really could cause physical damage."

If it's true that competition with friends kicks off physical stress responses, that's bad news, since we seem to engage in various forms of antagonism. A Western Michigan University study of competition and conflict in same-sex friendships found that while direct competition is more uncomfortable for women than for men, both sexes admit to trying to keep up with or outdo their pals. Men competed more in the areas of achievement (who makes more money, who knows more?), social attractiveness (who goes to more interesting parties?), controversy (the researchers' term for debating and arguing about events and values), and play (sports and games). Women were more competitive in the areas of affection (who is more caring?) and social skills (who is a better communicator/comforter, etc.?). Both sexes competed equally in the area of altruism (who is more giving?).

Fakers and Liars

"A true friend stabs you in the front," Oscar Wilde said. "True" friends by this definition may be rare indeed. Our close buddies, in fact, avoid extreme honesty in ways that could hurt us.

And they may even misperceive us because they are so invested in the relationship and therefore need to believe certain things about us so as not to rock its foundation. That's the conclusion that Weylin Sternglanz, Ph.D., and Bella DePaulo, Ph.D., came to after they found that while friends are better than strangers at using nonverbal cues to identify emotions, less close friends are better than closer friends at knowing when their friends are hiding negative ones. The idea that good friends are motivated to maintain a certain image of us echoes advice some psychologists dole out to couples: Having "positive illusions" about your beloved can hold the two of you together longer than a cold, clear view will.

One honesty killer is the need to believe a friend is exactly like you. Longtime friendship researcher and coach Jan Yager, Ph.D., calls it the "mirror image trap." "Does your friend fall into the trap of assuming everyone should approach life the same way that she does? Instead of respecting your differences, does she try to change you or tell you that you are 'wrong' even about issues or situations in which many opinions are equally valid?" she asks in her book *When Friendship Hurts*.

Dissecting the latest election or political scandal with buddies can set off personal conflicts as well as intellectual discord. It's easy to feel so passionately about an issue that you assume that all good and rational human beings, including your friends, naturally, will take your side. Some can relish a rousing debate and appreciate a friend with whom they can spar. But for those who themselves fall into the mirror image trap, a disagreement on a social issue can feel like a personal affront, one with a scary underlying message: "We're not really the same after all, and therefore, we're not as close and mutually admiring as we thought."

If you're on the other end of a debate with a pal who insists the two of you remain "twinsies," Yager writes, you might keep your opinions to yourself to avoid a confrontation. You'll just have to bite your tongue on charter schools or birth control or banking reforms or risk having a distinctly unproductive conversation that leaves you less enlightened about the issue at hand and more frustrated with your friendship. People caught in the mirror image trap might become excited about new friendships, thinking they've found their reflection, but then quickly sour on those burgeoning relationships once inevitable asymmetries appear.

Mark Vernon sees rampant dishonesty flowing through all friendships, not just problematic ones. "When you start to look," he writes, "it quickly becomes apparent that in a million little ways, as well as some large ones, friendship is often a matter of nothing less than faking it." But again, the motivation is purer than a desire to deceive. It comes instead from a fear that bluntness with a friend (yes, you have gotten wrinkly; no, you're really not talented enough to sing professionally) would destroy her. Sometimes lies of omission rather than commission are most commonly perpetrated: It's tricky to dodge the truth when you're directly asked for your opinion, but easy simply not to offer up your unsolicited critique of, say, a friend's handling of her difficult boss. And by the way, I've neither witnessed nor heard of any friend's taking a pastor's suggestion to "speak now" against a wedding in progress. We seem to "forever hold our peace," at least from the bride and groom, as a rule.

A lack of honesty toward friends, Vernon argues, can sometimes beneficially counteract our own tendency to misjudge. In other words, we're not always right anyway, so perhaps it's best that we spare our friends our "truth" since it might not be *the* truth. Slightly more malicious is our tendency to spill our buddies' confidences. Vernon quotes Nietzsche on this unpleasant feature of friendship: "There will be few who, when they are in want of matter for conversation, do not reveal the more secret affairs of their friends." It's fairly irresistible: You're at dinner, grasping for a topic of conversation, when gossip about a mutual friend springs to mind. A risk of intimacy with friends, it follows, is having your private trials shared with others.

Betrayals, Disappointments, and Demises

A friend who broadcasts our most sensitive secrets, however, is going beyond the human instinct to gossip and analyze private lives to learn where we stand in the world and what we should do to flourish. Like any close attachments, friendships have the power to hurt us profoundly. When cherished expectations are thwarted, or when people pull away unexpectedly or use their vast knowledge about us to create emotional weapons designed for maximum effectiveness, results can be scarring. Such experiences can sour our other relationships and make it more difficult for us to be positively influenced by friendships in the future.

Sixty-eight percent of those whom Jan Yager surveyed for her book had been betrayed by a friend. In addition to repeating confidences, common betrayals include spreading lies or rumors, stealing a lover, and not paying back financial loans. Betrayals are more common among friends who have grown apart. For that reason, Yager recommends letting those friendships die out rather than outright ending them, which could spark a desire for revenge.

Shedding friends naturally is quite common. Psychologist Laura Carstensen, Ph.D., mapped friend quantity over time and found that the number of people we hang out with dwindles after age seventeen, increases in our thirties, and declines again from age forty to fifty. Losing friends is inevitable. It would be very taxing to keep up with all the friends we've ever had and would prevent us from giving our attention to those who could best support us in a given time of life. Some sort of pruning as we travel along different stages only makes sense. Still, when two friends are not on the same timetable or of the same opinion about how close they should be with each other, egos get bruised.

Changes in marital status often disrupt friendships. "Over and over again, while researching my books, I heard the same complaint: My friends were always there for me, but then my spouse died and they don't seem to be comfortable around me anymore, especially those who are still married or in a relationship," Yager writes. "The same is true for those who get divorced: 'They've chosen sides,' I'm often told, usually by the one who is on the side that is being ignored and not chosen for continued socializing."

Parting ways with a buddy can be more difficult and complicated than breaking up with a boyfriend or girlfriend, says Susan Shapiro Barash, author of *Toxic Friends: The Antidote for Women Stuck in Complicated Friendships*. Though it might be naive, she says, we have a belief that friendships won't end, whereas most dating relationships are *expected* to end. Friend breakups, she adds, can challenge our sense of self, especially if we've been invested in and intertwined with a friend for many years.

The worst friend falling-out I've had was with a guy I'd

been close to for about six years. During that time, I was also close to his girlfriend, who later became his wife. I traveled about eight thousand miles to attend their wedding (putting the ticket on my credit card since I didn't actually have the money to pay for it), so you could say my commitment level to this friend was high. He'd introduced me to several wonderful pals of his, whom I had also gotten to know over the years. The actual "fight" was over something so minor that it's hard to recount without sounding ridiculous: He was in town, I met him for a drink to catch up, his work colleagues joined us, and I said I had to leave once it got late since I had to work the next day. He belligerently insisted I stay, saying he needed to talk about some problems he was having. I said I would have loved to talk about those issues earlier and that I hadn't known that his colleagues would be with us the whole time. We argued, and I left feeling stunned by his anger; it seemed to me that he had been unreasonable in wanting me to stay even longer than I already had.

I e-mailed him the next day, at the height of *my* anger, writing that if he didn't want to apologize, the friendship was over. He replied that in that case, the friendship was indeed over. I felt surprised, hurt at the loss of the friendship with him and, by extension, his wife, and frustrated that I wouldn't be able to control his version of what had happened. What if he made me out to his other friends to be the "bad guy"? And in flashes of self-doubt, I wondered if I had broken some unspoken friend code by not staying longer to listen to what was on his mind.

My bitterness and confusion faded into a disappointment that did diminish over time but occasionally bubbled up. I started to reevaluate the friendship in retrospect: Sure, this person had been a great support to me when I was heartbroken after a failed relationship. An adventurous soul, he'd exposed me to new places. But didn't others consider him to be arrogant—and me overly tolerant of his pretentious attitude? And didn't he drink a lot, meaning that I also did when in his company? Perhaps the incident, while small on the surface, represented larger differences that would have come to a head one way or another, eventually. And finally, wasn't his lack of willingness to apologize, or at least explain why he didn't think he should have to, a sign that he either didn't value the friendship very much or wanted out of it on some level?

Over time, I stopped thinking about him and our mutual friends (whom I "lost" in the breakup) on a regular basis, though I couldn't resist the occasional Google search. Then just last month I got a note in (where else?) my Facebook message in-box. "The details are a bit hazy but I know that I was a total jerk to you. For that I profusely apologize. I know the apology is like 6 or 7 years late and if you want nothing to do with me, I completely understand. I don't know how else to express it."

I was shocked and happy to get the apology and immediately accepted it. We made vague plans to get together the next time he's in town. It's a relief to have the proverbial thorn removed from my side. I've been just fine without him, but getting an apology, even seven years later, helps me appreciate the times we did spend together, rather than cast them in a dark light, as I had been doing. That arrogant, thoughtless lush? I can now restore him in my memory as a bon vivant who is worldly enough to have earned his habit of showing off occasionally. A friend breakup and the torrent of feelings it unleashes, I learned, can be more distressing than the actual day-to-day absence of a friend. A rift can make us question our judgment and whole chunks of our pasts.

More Fraught for Females?

Catfights are one thing, scalping your ex-BFF is quite another. Last year a nineteen-year-old New Mexican woman got into a fight with a former friend at a party and ripped off part of the other woman's scalp while pulling her hair. The incident is testament to the fact that former friends, once bearers of each other's intimate thoughts and feelings, can prove more hurtful to us than perpetual enemies.

A clear pattern emerged as I researched the dark side of friendship: It skewed female. Much of the research and almost all of the popular books and articles surrounding the topic of problematic friendship focus on the ladies. Are women harder to be friends with, born to be bad influences on each other? Or are they just more conscientious about wanting to actively ponder, learn about, and improve their friendships?

Kelly Valen conducted a survey of 3,020 women for her book *The Twisted Sisterhood*. Eighty-four percent of those respondents said they had "suffered palpable emotional wounding at the hands of other females. . . . Their prior experiences with other females have weighed them down, slowed them down and still tend to color their attitudes, relationships, self-esteem, and general approach to women." Seventy-six percent of respondents said "they've been hurt by episodes of jealousy and competition, while 74 percent have been stung by other women's criticisms and judgments and 72 percent have been the targets of gossip, rumors, and behind-the-back duplicity. At the same time, well over half admit that they too have been nasty and indelicate with other females."

This study has a few problems. For one thing, the respondents were a self-selected group, meaning that women who found the lengthy questionnaire at the local diner but hadn't had any female-friend traumas might have overlooked it. Secondly, Valen doesn't poll men to see if this is really a "sisterhood" problem or a humans being humans problem. And yet, curiously, a *Self* magazine/*Today* show online survey of eighteen thousand women came up with the exact same 84 percent figure for those claiming they've had a "toxic" friend who makes life difficult. (The survey came up with its own toxic taxonomy, including the Narcissist, the Chronic Downer, the Critic, the Underminer, and the Flake. If these types are so prevalent, the real question is: Which are *you*?)

Any psychologist would presumably agree with Valen's heartfelt message that no matter what the underlying causes of our rampant "relational aggression," as researchers call it, we need to consciously change our behavior and fight tendencies to inflict harm on friends (or would-be friends). Yet Peter DeScioli, Ph.D., and Robert Kurzban, Ph.D., who came up with the alliance hypothesis discussed in Chapter 2, would say that direct methods for curbing our very strong drive to gather allies and undermine those who would replace us in the "loyalty landscape" of our closest friends are unlikely to work. DeScioli and Kurzban write, "Research in this area has focused on its social harms, often regarding aggression as pathological. However, considering behavior in the context of strategy rather than pathology can be illuminating." Treating perpetrators as clever maneuverers and attempting to understand the rewards and costs of their behaviors in their particular social environment, rather than viewing them as "sociopaths," might just be a helpful shift.

"Men kick friendship around like a football, but it doesn't seem to crack. Women treat it like glass and it falls to pieces,"

Anne Morrow Lindbergh (aviator and wife of Charles Lindbergh) said. It's hard (not to mention dangerous) to generalize about why women might have more friendship conflicts than men, but Barash, who is a professor of gender studies at Marymount Manhattan College, says that on the basis of her research, "Relationships are so totalizing for women. For men, the bar isn't held as high, the expectations aren't as great, and the friendships aren't as significant." Furthermore, Barash says, women have been raised to think "there's not enough pie," perhaps because of their history of second-class citizenship. "So much becomes a rivalry because of how we are societally situated. There are 'Who looks best in this dress?' features in our popular magazines. Women are constantly compared to one another." Girlfriends can't help scanning one another's appearances and feel twinges of insecurity (or superiority) as a result.

Women often want what another has, Barash says—another reason that lifestyle changes might prompt friendship conflict: "One woman secretly can't wait to have children and live in the suburbs and her married friend says, 'She's still single and has her freedom.'" These common flashes of envy, which we are taught are wrong and make us bad, along with our fear of addressing conflict head-on, may explain why many are drawn to reality TV shows with over-the-top female fights. "We like to watch the mean girls," Barash says. "We identify and understand it, even though most of us don't want that kind of drama. These scenes represent our worst fears." I find that the closer I am to a friend, the less likely I am to feel envious when comparing myself to her, especially if it's on the superficial level of appearance, dress, etc. So it might be that it's among the general "peer group" of women (moms who parent differently than you do, for example) where the society-wide female judgment fest that Barash alludes to takes place. Reality TV stars, after all, are usually "peers" thrown together and not actual friends.

CLINICAL PSYCHOLOGIST TERRI APTER, PH.D., coauthor of *Best Friends: The Pleasures and Perils of Girls' and Women's Friendships,* traces back to childhood both the greater passion women feel for their friends and their higher expectations of them. In middle childhood and adolescence, she says, girls get to know themselves through their friendships. "They talk about themselves, they compare notes, they spot similarities that they couldn't have articulated with their family," she says. "They also learn that they have the power to entertain, to comfort, and to support someone else. Those are all good things to learn. But if you think that in order for a friendship to be viable it has to be pretty perfect, then when you see differences or when there is an inevitable change in the emotional weather or when you see that someone else prefers another friend and not you, you feel it's all coming apart. A lot of the real nastiness unfolds because girls don't know how to manage conflict. If something is a little bit bad, then they want to make it all bad. This is all one way in which the dark side of friendship is related to the positive side." Boys, on the other hand, Apter says, tend to have more of a group dynamic as children rather than intense one-on-ones. Adult male friendships are romanticized not for their intimacy level but for their easygoing camaraderie or loyalty in the face of adversity (note the plethora of war buddy films).

How might women bridge the gender gap in friend distress? "I think if women learned to take conflict out in the open and to see that it is perfectly consistent with friendship, then that would be a big improvement. Even middle-aged women often say they'll withdraw from a friendship rather than have an argument or express their dissatisfaction or disquiet," Apter says.

Women in the twilight years, in fact, are still avoiding friction with friends. An in-depth study by Robin Moremen, Ph.D., of twenty-six senior citizens found that most chose not to openly confront their friends when feeling hurt or disappointed. Again, the theme of thwarted expectations ran through these female subjects' responses to extensive questioning about the downsides of friendships. Strained ties were associated with the belief that friends will share similar interests and personal habits, be trustworthy, be honest, not exploit each other, live close by, not be overly dependent, share similar statuses, not be "whiney or demanding" when ill, maintain balance and reciprocity in their friendships, and tease each other only for fun. Friendship norms are highly valued but not often discussed outright.

Moremen is particularly concerned with the fallout from friendship norm violations among older women because friends are often more important to them than family members "and because negative events in their friendships are more detrimental to their well-being than positive events are beneficial. Older women run the risk of social isolation, depression, and material deprivation if rifts occur in their friendships."

One thought on the perceived challenges of female friendship is that just as individual women have historically and culturally been relegated to the "virgin/whore" dichotomy, female friendships are divided into Lifetime movie–caliber "laughter-through-tears" soul sisters, on one hand, and pairs of "toxic bitches" clawing each other and mud-wrestling, at least metaphorically, on the other hand. Too bad the truth—a full range of relationships among women—is harder to sell and grasp than those two stock portrayals.

So what do we know about the dark side of male friendship? Geoffrey Greif has found that men feel quite supported by their friendships, even if they don't tend to open up to their friends as much as woman do. They might spill their guts to the women in their lives and rely on buddies to provide a break from whatever is weighing on them.

I still have a hunch that younger men are less homophobic/macho and therefore more willing to examine feelings and struggles with each other; this might bring them more buddy drama but also more insight. On the other hand, women tired of reenacting their favorite soap operas could stop assiduously "maintaining" their friendships and take a cue from those poker-playing wisecrackers: Just enjoy each other's company—without imposing specific expectations on your friends or yourselves.

Here's one more research-based piece of advice for married women: Don't crash your husband's "guy's night." A Cornell University and University of Chicago study found that when a middle-aged man's partner becomes close to his male friends, he is 92 percent more likely to experience sexual dysfunction. That's not because he thinks his woman will have an affair or leave him for one of his friends, the researchers said. It's more that he derives his identity from his gang and feels his independence and virility threatened when his mate intrudes upon that psychological space.

From Mischief to Murder

Friends excel at coercing us to misbehave. Physical health is a not-so-obvious arena in which they may do so. The network research conducted by Christakis and Fowler showing that happiness and weight loss spread through friend networks also found that a friend who starts smoking increases your chances of taking up cigarettes by 36 percent. In a follow-up study on drinking, the researchers found that if your close friend is a heavy (i.e., daily) drinker, you're 50 percent more likely to drink heavily yourself.

We visualize friends who get trashed together as packs of young adults, but a Scottish study found that people aged thirty-five to fifty, while less socially disruptive when drunk, find it more difficult to say no to friends who push drinks on them than they did when they were younger. Those studied fessed up to lying about dieting or purposefully driving to events so that they wouldn't have to partake. It could be that these adults still want to appear young and carefree to each other, and for that reason they find it even harder to say no than they did when they were actually young and carefree. It's just one study, but it's an indicator that peer pressure to party is not at all an exclusively teenage phenomenon.

For those battling drug addictions, friends who still use can be a major impediment to recovery, says Carl Latkin, Ph.D., a professor at the Johns Hopkins Bloomberg School of Public Health. "A lot of our work is trying to assess someone's social environment. Which relationships within that environment are associated with a particular health behavior and then, how do you intervene to promote a behavioral change?" he says. In the poor sections of Baltimore, where Latkin runs programs and conducts research, people are more geographically isolated, meaning there's a smaller pool of potential friends. Latkin says this makes it harder to sever ties with those who are truly a bad influence. Making new friends might just be the key to the success of twelve-step meetings: "These programs seem to work because they give people something else to do and a new group to hang out with, a new forum for support."

It's often pointed out that drug users from disadvantaged communities such as inner-city Baltimore are punished more harshly in our criminal justice system than are white-collar criminals from gilded backgrounds. A recent case of negative friendfluence did land one high-finance fellow in jail, however. Raj Rajaratnam, a former hedge fund manager of more than seven billion dollars in assets, was convicted on fourteen counts of securities fraud and conspiracy. Essentially, he illegally traded on tips and info he pulled out of several friends. As the *New York Times* put it, "He fed the needs of those in his orbit who could be helpful to him, whether with money, tips, or friendship. In his soft-spoken manner, shaped by his years at secondary school and college in England, Mr. Rajaratnam alternately prodded, chided, ridiculed and flattered his sources. Above all, he was a good listener, saying little as those on the other end of the phone, eager to impress the hedge fund titan, kept talking."

A particularly sad anecdote involves Rajaratnam's old business school buddy Rajiv Goel, who had plateaued as a midlevel manager at Intel while his friend soared among the 1 percent of the 1 percenters. The two vacationed together and talked frequently; recorded conversations obtained by the government reveal that Goel was desperate for his friend's approval. He went so far as to share confidential information about

Intel's earnings and investments. The *Times* recounts a piece of the trial. "'He was a good man to me,'" said Mr. Goel, who pleaded guilty and took the stand on behalf of the government. "'I was a good pal, a good person to him, so I gave him the information.'" In return, Rajaratnam lent him six hundred thousand dollars and made him even more in the stock market. While Goel should hardly be let off the hook, his willingness to put himself on the line is testament to the strength of human ties over legal restrictions.

Sometimes friendship produces a sum worse than its parts. In 2001, when they were both young adults in the D.C. area, Clara Schwartz and Kyle Hulbert became comrades for the same reason most people do: They had the opportunity to meet (at a Renaissance fair), and they shared interests (magic and fantasy).

It sounds like a benign, if nerdy, meeting of minds, but their alliance soon took a turn toward the sinister, in part because they had something else in common: mental instability. Clara, who had bouts of paranoid thinking, was incensed with her father, with whom she had a rocky relationship. Kyle, who had symptoms of schizophrenia, was delighted to have a new friend and, in a reenactment of his extensive "warrior" fantasies, began to see himself as Clara's honorable protector. Which is why after she asked him to kill her father, he drove to the farmhouse where the older man lived and stabbed him to death with a twenty-seven-inch sword. Many factors led to Clara's father's death, but he would likely be alive today if not for the friendship between Clara and Kyle. (Both are in prison now, and Kyle has a profile up on cellpals.com, a Web site for inmates looking for pen pal correspondence. "What I am seeking is very simple: I'm looking for friends," reads his post. "With sixty-four million people on the Web, I think my chances of finding friends are rather high, wouldn't you agree?")

The Curse of the Similarity Drive?

We're drawn to potential friends who are similar to us in terms of beliefs and demographics. Could these friendship pairings, en masse, be a bad influence on society at large? In contemplating the growing class divide among Americans, the *New York Times* columnist David Brooks describes an "upper tribe" concentrated in coastal cities, whose members are self-disciplined, productive, and socially conservative—that is, whether or not they have liberal political views on paper, they are personally less likely to have children out of wedlock or to divorce than low-income people who espouse more conservative beliefs. Less educated folks in the "lower tribe" live in "disorganized, postmodern neighborhoods in which it is much harder to be self-disciplined and productive." Brooks wonders if a national service program forcing members of these two tribes to live together would "spread out the values, practices and institutions that lead to achievement. . . . If we could jam the tribes together, we'd have a better elite and a better mass."

Dalton Conley, the NYU sociologist, imagines more complicated consequences of tribe jamming. *Friends with Money* is not just a Jennifer Aniston film; it's something many of us have struggled with. Having more or less than your friends is a difference that must be constantly negotiated. It's not fair for the wealthier friend to suggest expensive outings or give lavish presents that can't be reciprocated, but perhaps neither is it fair for the poorer pal to deny a friend the perks of a lifestyle she can afford, out of a need to make things superficially "equal."

"Modern friendship is supposed to be distinct from the market sphere," Conley says, "but if you have inequalities between folks in the friendship, sometimes they become so great that they intrude the market onto the friendship and it becomes very difficult and awkward."

More concretely, Conley recalls again the wealthy family that broadened his horizons when he was a project-dwelling teen in New York: "They were incredibly generous to me. But I remember many times when, for example, just the two kids of the family and I would go to get lunch and I would be thinking, 'I've only got five dollars. I need to be careful.' One of the brothers would have a twenty-dollar bill in his pocket, and would order three Cokes for himself. And then the bill would come, and he'd say, 'Let's just split it.' This caused an incredible amount of anxiety for me. There are certain limits to cross-class friendships." People who move up socioeconomically as adults, Conley points out, have to pivot and live in two worlds—that of their family and that of their new friends in a higher class. Especially for wealthy minorities, the dichotomy between one's social life and kin group can be hard to manage.

Friendships across different socioeconomic groups have their challenges, but so does sticking too close to your tribe. Self-selection can serve to overemphasize the traits and beliefs people have in common, skewing their perception of the world at large. A beauty maintenance example springs to mind: How many women in Hollywood shell out money for lip and cheek implants, for instance, when these alterations strike many of us outside that world as horribly unattractive? To those constantly comparing themselves to people with that same look, it's apparently a desirable and aesthetically pleasing one.

C. S. Lewis considers the curse of similarity in *The Four Loves*. He writes that every real friendship is

> a sort of secession, even a rebellion. It may be a rebellion of serious thinkers against clap-trap or of faddists against accepted good sense; of real artists against popular ugliness or of charlatans against civilized taste; of good men against the badness of society or of bad men against its goodness. . . . The danger is that this partial indifference or deafness to outside opinion, justified and necessary though it is, may lead to a wholesale indifference or deafness. . . . Like an aristocracy, it can create around it a vacuum across which no voice will carry.

It initially sounds alluring: two friends locked in their own private world, where their shared views reign and where they can always feel smug and superior to all others. And yet reveling in similarity isolates people from others and eventually causes friends to metaphorically suffocate from a lack of fresh air from outside.

We know that preserving the cohesiveness of a group can trump good judgment; classic work on "groupthink" explains how clusters of like-minded people can make bad decisions by ignoring alternatives and even dehumanizing other groups. (Some have argued that the 2003 Iraqi invasion is an example.) And the large majority of hate crimes are committed in groups of four or more. The greater the pack of friends involved in such crimes, the more brutal they are.

Even the seemingly virtuous practice of valuing and caring for friends can trigger damaging consequences in the context of an unstable society. Daniel Hruschka describes how friends are more important in places where "uncertainty is pervasive, in the form of food scarcity or inaccessibility due to high prices, material unavailability, violence, or lack of legal sanctions." But while friends can compensate for structural flaws in a country, they can ironically exacerbate them: "During the Soviet period in Russia, for example, friends not only provided a stopgap for a failing centralized economy, they also helped each other beat the system, thus further weakening any system that might have existed. . . . Thus, a reliance on friends can lead to a vicious feedback cycle, in which friends become more and more valuable as broader social institutions break down."

WE CAN TRY TO BEFRIEND ANYONE we want, yet counterintuitively, having too large a pool of potential friends can lead to more predictable yet less satisfying relationships. You'd think students at big, diverse colleges would have more diverse friendship groups than do those at small, more homogeneous schools, but Angela Bahns, Ph.D., an assistant professor of psychology at Wellesley College, found just the opposite: Pairs of friends were significantly more similar in terms of attitudes, beliefs, and health behaviors on a large campus than at the smaller colleges included in her study.

Bahns reasons that people from the bigger schools were able to satisfy their similarity drive more thoroughly than those from the smaller schools. "They took a more fine-grained approach in finding a closer match," she says, "whereas in smaller environments, people might have that same fundamental drive but are constrained by the environment, meaning their criteria have to be relaxed." It's not simply that in the smaller environments students befriended others with different hobbies or interests. It's that they had friends with different outlooks and lifestyles, which presumably opened their minds in the fashion that college is supposed to.

Secondly, and more surprisingly, the friendships on the smaller campuses, among people who were less similar to each other, were closer. "My guess is that the higher quality of these friendships stems from the fact that the students are aware that they are kind of stuck with who they've got, and if a friendship doesn't go well, they can't just trade that friend up for a better one." (This is the inverse of what many have said of online dating: It creates a sense of endless possibility and therefore less of an urge to commit to just one.) "A lot of research implies that because it's an attraction, similarity is a good thing that leads to longer, more satisfying relationships. Here we have a little bit of data that questions that," Bahns says.

The theory that having more choices in friendship selection isn't necessarily better is corroborated by some international work Bahns has done showing that in Korea, for example, where a more interdependent mode of thinking leads people to pick friends who are already embedded in their social networks (rather than search outside for pals who more closely reflect them as individuals), friendships are just as strong as Western ones.

As much as we tend to befriend people who are similar to us, we sure do seem to love unusual animal pairings. The Internet is lately ablaze with cute videos of a hippo cuddling with his turtle pal, or a chicken roughhousing with his dog buddy. A human equivalent of such odd couples exemplifies the mix of bad and good influences unlikely friendships can contain: Since the 1970s, Cuban dictator Fidel Castro has been close friends with the Nobel Prize–winning novelist Gabriel García Márquez. The book *Fidel & Gabo* by Angel Esteban and Stephanie Panichelli-Batalla investigates their apparently deep friendship. Márquez even claimed once that "Fidel is the sweetest man I know," and he shows Castro all his manuscripts before he sends them to his publisher. (The Comandante is an ace proofreader.)

"The first phase of this relationship was really a friendship of interest for both," says Panichelli-Batalla, an assistant professor of Spanish at Aston University in Great Britain. Fidel was happy to have the support of an internationally revered intellectual such as Márquez. "In the case of Márquez," she says, "we concluded that he had a sort of obsession with power. Befriending Fidel was a way of having some kind of power without being a power figure, since he wasn't interested in that. But then at some point, it became a real friendship."

The costs of this relationship to Márquez include a falling-out with another close friend, the fellow novelist Mario Vargas Llosa, who, like most left-leaning Latin Americans, turned against the Cuban Revolution after initially supporting the idea of it. (One benefit of the friendship with Castro includes a mansion on the island for Márquez when he and his family visit.) The book notes instances of friend-fueled bias in Márquez's generally objective journalism: He reported the Elián González affair in a nonnuanced, pro-Cuba manner, for instance. But the authors also assert that Márquez does take Castro to task on certain policies, even though he never speaks against him publicly. "He's managed to have conversations with him and to criticize him," Panichelli-Batalla says. "He is perhaps the only critic Castro will listen to. Castro really respects him." Castro even released prisoners at Márquez's behest. Fulfilling a favor for a friend can supersede even the hardest dictator's methods.

AS COMPLEX AND CUTTING AS friendships can be, the worst influence of all is perhaps a total lack of friendfluence. Social neuroscientist John Cacioppo, Ph.D., describes loneliness as the fallout of not fulfilling a biological need for social contact, one almost as strong as thirst or hunger. Thanks to his work, we know that loneliness is associated with the progression of disease, alcoholism, and suicidal ideation. Exposing yourself to the vulnerabilities that come with intimacy, withstanding the discomfort of connecting to people who are different from you, and striving to assert your true self among pals eager to sway you in other directions are most certainly worthwhile struggles compared to the bleak alternative of friendlessness.

CHAPTER 7

Screenmates: How Technology Affects Friendship

STRANGE AS IT SOUNDS, TONI BERNHARD doesn't know if one of her closest friends is dead or alive. Marilyn lives (or lived) in Sydney, Australia, about seven thousand miles from Toni's home in Davis, California. Toni knew all about Marilyn's ongoing treatment for late-stage breast cancer, and suddenly, after a year and a half of daily correspondence, Marilyn's name stopped popping up in Toni's in-box. She could only expect the worst. "I felt completely cut off," Toni says. "I don't know Marilyn's husband; I don't even know if he knows how close we are. I have no other way of getting in touch with her. I went from being in constant contact with someone I just adored to a complete blackout."

Toni, sixty-six, is virtually bedridden and conducts most of her friendships via e-mail. After contracting what she and her husband jokingly called "the Paris flu"—she caught it while they were on a long-anticipated trip to France in 2001—Toni, astoundingly, did not recover. "My immune system never went back to normal. It's like having the flu all of the time, or an extreme feeling of jet lag. The doctors call it 'chronic immune system activation.' They're hoping that something will happen to reset it. I'm always trying antivirals, acupuncture, and Western and Eastern modalities. So far, nothing has worked."

A former professor at the University of California, Davis, School of Law, Toni had to quit her beloved job and scale her life down to fit the limitations her body imposed. She fell into a depression. Eventually, she began reading blogs by and for chronic illness sufferers. A witty comment or thoughtful observation occasionally compelled her to make direct contact with an intriguing fellow poster. Some of those exchanges mushroomed into friendships. Then Toni, who has studied Buddhism since 1992, began to write a book, *How to Be Sick: A Buddhist-Inspired Guide for the Chronically Ill and Their Caregivers*, about her accepting stance toward her illness. She started a Facebook page to promote the endeavor and watched with delight as it sprouted an active community of people dealing with chronic illness. It was the book, in fact, that first prompted Marilyn to e-mail her.

Toni currently has four other close Internet friends. They are all women with chronic illnesses, but that circumstance no longer dominates their exchanges. A few have been in daily contact with Toni for eight years now. "My friends used to be people here in Davis, and coworkers. I don't have much in common with my former friends anymore. They would do anything for me if I called them, but we don't have, for example, the students to talk about or complain about anymore." She does touch base with two "3-D" friends on a weekly basis: Richard, whom she's known since college, and Dawn, whose kids grew up with Toni's. If she's feeling well, she meets Dawn at a coffee shop; otherwise, Dawn, a realtor, stops by the house to chat.

"It's interesting," Toni muses. "My relationship with Dawn is richer because we've shared more experiences and know each other's kids. But with my friend in New Hampshire, we write things like 'I'm holding you in my heart' and 'Dearest, how are you?' or even 'I love you.' I would never talk to Dawn that way! It would be totally out of character. I think you're willing to put things in writing that you don't say face-to-face. I'm tempted to say that I'm closer to Dawn because I can call on her if I need her. But these other friends are people I e-mail either every day or every other day. With Dawn or Richard, we don't say, 'What happened on Monday?' One of us raises some subject, and we discuss that. It means I know less about their day-to-day lives than those of my Internet friends."

But couldn't it be that Toni, who doesn't even video chat via Skype with these friends because it tires her and she finds it awkward, is simply projecting qualities onto these disembodied souls with whom she claims such communion? "I might have felt that way in the first few months of knowing them. But all these years later, I've seen all sides. I've seen their crankiness." One friend, Laura, initially conveyed a Pollyannaish tone in her e-mails that made Toni doubt whether they could ever truly become close, even given Toni's efforts to cultivate a "Zen" attitude. "She had a positive gloss on everything—you know, 'God has a reason.' Then she had an MS flare-up, and she started to be more honest with me. At first she was apologetic and would write, 'It's not the right way to feel.' I would write back: 'But it *is* how you feel!' We broke through the facade.

"I sometimes complain to my husband that I love my Internet friends, but it's not as satisfying as being with someone in the flesh that you can touch. And there are just so many nuances that don't come through in an e-mail." Since she's been married for four decades, it's also difficult for Toni to have friends her husband doesn't know at all. But the complementary aspect of having Richard and Dawn, on one hand, and her electronic pals on the other, fulfills Toni's friendship needs.

Toni is even grateful to have come down with her debilitating condition at this point in history. "If you're going to get sick, it's good to do so in the Internet age," she says. "Many women with chronic illnesses tell me that the Internet is their savior." To say the loss of her friendship with Marilyn is less painful than it would be had they known each other in person is to underestimate the intensity of feeling that can run through global data streams. But it would, at the same time, underestimate the benefits of the shared ties that come with more traditionally formed friendships. Because the friendship existed outside of their respective social networks, Toni likely won't get a phone call from someone who knew Marilyn and who can grieve along with her, helping her through that process. "Marilyn was someone I had this wonderfully silly friendship with," Toni says with audible sadness. "She would play games on her iPad. I would, too, and we would take screen shots of our scores and such. I'm starting to realize that she is gone from me."

I probably don't need to tell you that the way we go about the business of friendship has fundamentally changed since the rise of the Internet twenty years ago. (Friendship has even become the basis of many actual Internet businesses during that time.) The very use of the word "friend" has expanded; who would have foreseen ten years ago that "to friend" would be a verb? Yet the most familiar comments to arise in discussions on the intersection of technology and friendship are along the lines of "I hate to break it to you, but your one thousand Facebook friends are not your friends." I think even social media novices distinguish between "friends" as shorthand for online contacts and true friends. Nonetheless, in today's media environment, real friends are commingling with other people who are less than friends more than ever before, on our screens and, perhaps, in our consciousness.

Many social scientists and cultural critics have suggested that our involvement in cyberspace is harming our real-life interactions, rotting our brains, and diminishing our very abilities to be human. Rapid change spurs fears, and there are certainly some things to be concerned about, in terms of the unknown long-term effects of our screen-crazy lifestyles. But sociologist Keith Hampton, Ph.D., associate professor in the department of communication at Rutgers University and a leading expert on technology and social life, argues that the dangers are exaggerated. A fear of new technologies and romanticization of the past, he says, characterize the history of sociology. "From electricity to the bicycle to the telephone, sociologists have argued that technological advances cause a decline in family and community life," he says. "These arguments have kept social scientists in business all these years." In the '90s, sociologist Robert Putnam argued in *Bowling Alone* that television was replacing the fun civic activities of the '60s. But during that heralded decade of the '60s, another sociologist, Robert Nisbet, was busy waxing nostalgic about the '30s, an era he viewed as high on social connectivity.

While it's too early to predict their specific long-term effects, Hampton says, we know that social media sites have made friendships more persistent (we no longer give up most ties as we move into new phases of life) and more pervasive (we have twenty-four-hour access to incoming information about the people we know). "The criticism of the pervasiveness of social contact is that these tidbits are meaningless, like little bits of fluff," says Hampton. "But maybe the fluff accumulates into something substantive." We also know that people who use social media more also socialize more in traditional settings such as cafés. Thus, they have more diverse overall social networks than do lighter users or nonusers of social media. "The image of the social media user as the guy alone in the basement in his underwear is not the reality."

You now have the power to befriend a wider variety of people, and to maintain those friendships, more easily than anyone has been able to throughout human existence. Being tethered to electronic devices while you interact with those friends, however, can cause a distinct yet related set of problems. As we sort through the good and bad psychological effects of texting, IMing, tweeting, posting, e-mailing, and so forth, what arises as most interesting is the extent to which our online behaviors align with our off-line behaviors and even with our most primitive instincts. (Consider the number of fellow Twitter users people tend to interact with before becoming overwhelmed: It's 150—that same old "Dunbar's number" that refers to the average size of the tribes we lived in when human brains evolved.) After all, the basic urge to connect with others predates the urge to connect with them online by tens of thousands of years.

How We're Friending

First there were e-mail and online message boards, and then the original social networking Web site, Friendster, which began in 2002. Friendster petered out as MySpace, which catered to aspiring musicians and youth culture in general, took over. Finally, after it began in 2004, Facebook fanned out from its base of Harvard students and is now the dominant social networking site. (Google Plus hasn't really taken off; I'm sure some other site will rise to prominence by the time this sentence is printed.) Concurrently, Twitter and LinkedIn have become top social network destinations, though LinkedIn is primarily for professional contacts, while Twitter mingles your favorite friends, relatives, potential dates, potential employers, and A-, B-, and C-list celebrities into one cacophonous conversational stream. As of late 2011, 65 percent of adult Internet users were on a social networking site, up from just 5 percent in 2005. Eighty-three percent of Internet users ages eighteen to twenty-nine and a full 33 percent of users over sixty-five were on networking sites. Those surveyed described their experiences in mostly positive terms such as "fun," "great," and "convenient."

Lone bowlers, take note: Internet-using Americans are more likely than others to be in a voluntary group or organization (80 percent, versus 56 percent of non-Internet users). Such group participation is even stronger among social network users (82 percent) and Twitter users (85 percent). Overall, 25 percent of Americans are not in any of the groups researchers studied, yet it's clear that those who are online are more engaged in various community activities than those who are not online.

Do your male Facebook friends post more pundit diatribes while your female ones tend toward baby picture displays? University of Texas, Austin, researchers found that men are a bit more likely to post news and information and share political and religious views, while women are more apt to express affection and post family photos. (Eighty-four percent of both men and women showcase their "relationship status" on the site.)

If you're on the site yourself, you might have noticed that your Facebook friends have more friends listed than you do. In 2012, the average user had 245 friends, while the average friend had 359 friends. This is the online version of a sociological principle called the friendship paradox: You're more likely to be friends with popular people than less popular people. Keith Hampton uncovered the existence of "power users" on Facebook. Making up 20 to 30 percent of users, they account for the fact that the average Facebooker gets more messages, requests, and photo tags than he or she gives. The power users come in different flavors: Some are liberal with their "likes," some collect a lot of friends, and others tag a lot of photos, making sure their friends are identified wherever possible (regardless of how photogenic they happen to be, presumably). Hampton points out that power users could take on the role temporarily: After your vacation to Bali, you might be a power photo poster, and while sick in bed, you might become a power "liker."

Now that a critical mass is using social media sites, the latest trend is "pruning": Sixty-three percent of users have deleted "friends," compared with 56 percent in 2009, while 44 percent have deleted comments by others on their profile, and 37 percent have untagged themselves from photos. Women are more likely to "unfriend" than men. Of course, this isn't the equivalent of an actual friend breakup since castoffs could be perfect strangers or acquaintances who post too many stupid videos. Pamela Rutledge, Ph.D., the director of the Media Psychology Research Center, says she's heard claims that this is an indication that people are pulling back on Facebook and Twitter, but she sees it as a natural response to an initial friending fervor: "I noticed that my Twitter feed has become junked up with people tweeting about things I don't care about or in languages I don't understand. I've cut down on those I follow, because I've figured out what the tools can do."

This kind of tailoring and trimming of one's online social contacts is a fascinating counterpoint to the social world we live in off-line, which is somewhat circumstantial (which friends do we work with or live near?) and somewhat the product of desire and effort (which friends do we actively make plans with and travel to see?). Social media and even e-mail programs allow for an easy "accounting" of all those we know. Once circumstance and mutual effort are removed from the equation, which buddies do we interact with the most? Which are we most influenced by? Sometimes the answer depends on yet another factor: the frequency of *other* people's online activity.

My L.A.-dwelling friend Erika posts almost daily about her work and travel adventures (and the antics of her cat, Boris, which sounds boring but somehow isn't—chalk it up to friend-love blindness). She therefore distinguishes herself and pops up, literally, from the "flat" structure of my online friend landscape (as Pamela Rutledge describes the nonhierarchical nature of online social life). As such, the updates and pictures she posts worm their way into my consciousness, even though I live three thousand miles away: Her yoga practice makes me aspire to be healthier (I admit it hasn't yet triggered action in that direction, but it's priming me on some level, I just know it). Her career success reminds me that I can do more and be more (in addition to being a role model from afar in the work arena, she actively encouraged me when I mentioned I wanted to write a book). And her willingness to try new things (surfing is just the latest) gets my imagination going in the direction of endeavors I've wanted to try but haven't got to yet (like acting classes). If she, like some of my other friends, were not so pro-Facebook, those same qualities of hers would still influence and inspire me, but not as regularly since she lives so far away. Consistent power users, take heed. Recognize the other power that comes with your penchant for posting: You are a louder and possibly stronger influence on your more Facebook-shy friends. Use it for good and not for evil.

Clay Shirky, Internet and society guru and professor at NYU, argues that we need a new way to conceive of the murky term "media." Here's what he proposes in his book *Cognitive Surplus*: "Media is the connective tissue of society. . . . Media is how you know what's happening in Tehran, who's in charge in Tegucigalpa, or the price of tea in China. Media is how you know what your colleague named her baby. Media is how you know why Kierkegaard disagreed with Hegel. Media is how you know where your next meeting is. Media is how you know about anything more than ten yards away. All these things used to be separated into public media (like visual or print communications made by a small group of professionals) and personal media (like letters and phone calls made by ordinary citizens). Now these two modes have fused."

Your friends are no longer a cordoned-off refuge from the "real" world. Some of them might be people you don't ever *see* in that so-called real world off-line. Instead, they are flowing alongside the constant input of messages and information, with your work stresses, world crises reports, television trivia, your mom's funny video recommendations, your girlfriend's love notes, and so forth. Is friendship adulterated by this data tsunami? Or are we generally better off by having friendly avatars, handles, usernames, and thumbnail photos there in the torrent?

Gloomy Messages

In an essay where he argues that friendship has become so broad and universal that it is now meaningless, the literary critic William Deresiewicz, Ph.D., points to social media as emblematic of friendship's demise. "Facebook's very premise—and promise—is that it makes our friendship circles visible," he writes. "There they are, my friends, all in the same place. Except, of course, they're not in the same place, or rather, they're not my friends. They're simulacra of my friends, little dehydrated packets of images and information, no more my friends than a set of baseball cards is the New York Mets." He goes on: "Friendship is devolving, in other words, from a relationship to a feeling—from something people share to something each of us hugs privately to ourselves in the loneliness of our electronic caves, rearranging the tokens of connection like a lonely child playing with dolls."

Facebook critics will light up with recognition at Deresiewicz's sharp descriptions of the absurdities and narcissism inherent in social media practices. Before, we used to tailor what we said and how we expressed thoughts to a specific friend or group of friends. "Now we're just broadcasting our stream of consciousness . . . to all 500 of our friends at once, hoping that someone, anyone, will confirm our existence by answering back," Deresiewicz writes. Before, we would have called a friend in a city we planned to visit to personally invite him out. Now we might write on his "wall" something like "Coming into town next week. Can you meet me for a drink?"

Deresiewicz finds the exhibitionism of private messages for a public audience particularly disturbing. "There's something faintly obscene about performing that intimacy in front of everyone you know, as if its real purpose were to show what a deep person you are. Are we really so hungry for validation? So desperate to prove we have friends? . . . And the whole theatrical quality of the business, the sense that my friends are doing their best to impersonate themselves, only makes it worse. The person I read about, I cannot help feeling, is not quite the person I know."

Finally, Deresiewicz rails against how Facebook reduces us to a list of random interests and decontextualized information. What forges real friendship, he argues, are shared experiences that allow people's stories to come out. Through those stories, a friend's character traits—qualities you surely can't describe in a tweet or update—are revealed, over time.

I couldn't agree more about how solid friendships develop. That's why many strong ones unfold during high school and college, a time when people have more hours to spend simply talking and hanging out, without a higher purpose other than to enjoy each other's company and learn about each other's histories and ways of thinking and feeling. There is a startling ring of truth to what Deresiewicz says; many of us have felt annoyed at friends' attention-seeking or self-promotional behavior online, even perhaps while we ourselves are committing the sins of bragging or soliciting approval for life choices via a not-so-Greek chorus of "likes." Yet his implied premise seems to be that we've altogether stopped those storytelling conversations and character-revealing enterprises with friends. In fact, most of us use social media to keep tabs on friends we've already made and met in person or to stay abreast of the comings and goings of a hodgepodge of acquaintances and old pals. In either case, we're simultaneously spending one-on-one time with close friends, geography permitting.

I've spent only one weekend with Sofia in the past two years; there's no question that just being with her over those two days was more satisfying than any one Skype session, e-mail exchange, or Facebook posting we've shared since then. But still, the cumulative weight of those electronic tête-à-têtes keeps our friendship fresh despite our physical distance. I simply like (in the traditional sense, as well as the Facebookian sense of the word) to see pictures of her daughter hamming it up or to read her take on the legislative debacle that has her incensed of late. I'm grateful that I can send a quick e-mail to her about a concern, without having to sit down and craft a long handwritten letter to justify the act of snail-mailing. (I'm even more grateful when I get a quick and comforting response.) I'd rather be able to leisurely share stories like we did in the old days, but that's tantamount to saying I'd rather be nineteen again. In light of life's harsh realities, virtual reality is pretty good. It's not a full substitute for friendship, but Facebook is not destroying our ties; it's helping us maintain them.

Even if we're not in as dark an era for friendship as Deresiewicz describes, there are definite downsides to palling around online. Philosopher Mark Vernon calls one drawback "the casual callousness of the internet." He writes, "Most people will have had an email that, written in a hurry, struck them more forcefully than was intended. These short messages, which now rule millions of lives, are misunderstandings waiting to happen." And sometimes, the fantasy of a synchronized rapport is shattered by the realization of how one-way electronic expression can be: "You can be pouring your heart out into a chat, while the individual on the other side of the screen is yawning with boredom, gossiping with someone else, or suddenly decides to switch off."

We still don't have universally accepted rules for online behavior; this creates old-fashioned etiquette problems and attendant hurt feelings. Henry Alford, author of *"Would It Kill You to Stop Doing That?" A Modern Guide to Manners*, says, "The essential trouble with e-mail is its dead-fish aspect. If you say 'Sure' aloud, it sounds positive, confident, full speed ahead. But if you write an e-mail response that simply reads 'Sure,' that doesn't look so, uh, sure. You'd do better with 'Sure!' or 'You bet' or 'Absolutely.' You have to overcompensate for e-mail's blank affect. Exclamation points are your very best friend."

The intuitive way that we navigate this friend etiquette Wild West seems to be by measuring friends' past online behaviors against their current ones and reacting to aberrations. "So-and-so always e-mails me back right away. I wonder if she's mad at me." Or: "So-and-so never calls me, so I wouldn't think to take offense at him posting a birthday message online instead of picking up the phone."

Andrea Bonior, Ph.D., clinical psychologist and the author of *The Friendship Fix: The Complete Guide to Choosing, Losing, and Keeping Up with Your Friends*, is on the front lines of online friendship conflicts. The number one casualty of excessive online use is, naturally, time that could be better spent. "I've talked to clients who are busy at their jobs but really lonely in the town they live in; they don't have a friend to go to brunch with or to join them on a run," Bonior says. "But if you get into the nitty-gritty of why they're busy, they are crashing on the couch after work and searching the Internet all evening. They are clicking 'like' on stuff when they should be using that time to join a book or knitting club or to go to the dog run or a park where they can meet other people."

Bonior has seen shy clients who get so used to connecting online that face-to-face interactions—the very kind they need to practice to keep their anxiety at bay—become riskier and feel less comfortable compared to an e-mail they can craft at a distance. "It's a self-perpetuating cycle," she says. "That's when people can really start to suffer, because we do need to feel like we're a part of the 'outside' world. I think the art of spontaneous conversation is being lost because we're focused on controlling our communications, instead of going along with the give-and-take and stumbles that come with face-to-face interaction."

Those with low self-esteem might not benefit as much as others from social media engagement. A study from the University of Waterloo found that people who don't think well of themselves are particularly attracted to the notion of posting comments online and getting instant feedback from a large group of people. However, when they post such comments, they tend to emphasize struggles or negative feelings ("just lost my keys" or "feeling ugly today") and therefore end up alienating their Facebook friends, which in turn confirms their low assessment of themselves. In contrast, posters who put up cheery messages were rated as more likable. In what seems to be yet another instance of the Matthew effect at work on friendship, those rich in self-esteem get it boosted further by public approval on social media sites, while those poor in it get the opposite of the support and affirmation that they're attempting to solicit.

(Incidentally, a recent study found a link between certain narcissistic personality traits—namely, "entitlement/

exploitativeness" and "grandiose exhibitionism"—and the total number of Facebook friends. I wouldn't go around diagnosing your friends on this metric alone, but it raises interesting questions about personality differences and stances toward friendship: Do more showy types see friends as numbers to rack up, while modest people might cringe at the virtual flash and noise of Facebook as a social environment?)

Facebook in particular can give rise to an unhealthy and misguided surge of envy. Sociologists Hui-Tzu Grace Chou, Ph.D., and Nicholas Edge, of Utah Valley University, asked 425 students about their Facebook activity level and whether they agreed with such sentiments as "Many of my friends have a better life than me." It turns out that the more time the subjects spent on Facebook, the more likely they were to believe that other people were happier and better off than they were. The correlation grew stronger for students whose Facebook "friends" included a greater proportion of people whom they didn't actually know. Finally, students who socialized more in the real world than they did in cyberspace were less likely to express general unhappiness.

Bonior sees envy crop up most often when her clients' friends are in the midst of life transitions. "I work with a lot of young women," she says. "And when some of them see things like marriage and pregnancy or home purchases announced on Facebook, they feel a stronger sense that something is missing in their own lives." If a client who has been through fertility treatments stumbles upon picture after picture of her friend's adorable newborn, or if someone else who longs for a long-term love is bombarded with daily updates on a friend's elaborate wedding planning, it's easy for her to forget that all lives are complicated—even in the midst of seemingly joyous events.

"In the past," says Bonior, "you could say, you know, I'm going to decline this baby shower invitation because I'm too sad to be around a pregnant woman. Now you have to block somebody from your newsfeed."

Moreover, the social pressure to comment positively on such edited montages of bliss makes envious or dejected women express the opposite of what they're really feeling. "They think, 'Well, twenty-seven people have already liked this photo, so I guess I should chime in.' Then they'll comment, 'Oh, how beautiful!' or 'I'm so excited for you!'" Bonior says. The poster herself has no idea who is and isn't comfortable with her onslaught of cheer. "Back in the day, somebody would present a slide show from their Tahitian honeymoon and they could see when peoples' eyes were glazed over. They could take in the cues and wrap it up."

What has surprised Bonior most as she has tended to the emotional wounds of frienders in cyberspace is the sometimes devastating impact of online communities gone awry. A typical example would be a group of people who "met" on a photography message board and then became close to one another, ramping up their posts and group communications. Then the group breaks down as members employ middle school–style exclusionary tactics, such as blocking someone or starting a new group and not telling some outcast members. A client might tell Bonior, "This group was my life. I was online with them all day for two years, and all of a sudden they turned on me because they don't like what I said to one member, and now I feel completely ousted."

It's a much starker loss, Bonior points out, compared to what might have happened in a social circle twenty years ago, where one would spin out of a group more gradually and retain ties to some members. "You'd never guess that with a click of a mouse a life could come crumbling down, but it shows you how much meaning these relationships have. At three a.m, you can log in and someone else from the message board or listserv can be right there with you." What's worse is that while these particular patients feel completely bereft in the wake of mean girl–style cliquishness, Bonior says, they also feel silly for feeling bereft. " 'I never really knew these people,' they might say. But sometimes they knew them better than most of their closest real-life friends."

WE KNOW WHAT ONLINE LIFE can do to our hearts, but what about our brains? I've never heard of a friend addict. But the diagnosis of addiction, so liberally assigned to behaviors these days, has been applied to the use of social media sites and the Internet in general, of course. Vernon points out that in South Korea, "probably the most connected country on the planet, the number of people who have died from blood clots caused by sitting too long at a terminal is now counted in the dozens, and it is reckoned that 210,000 children have an addiction problem." Social media might be more "addictive" in that sense because the cost is so little compared to the cost of drinking or smoking if one is hooked on those behaviors. But constantly giving in to an urge to check these sites does disrupt your schedule and waste your time.

An Italian team recorded different physiological reactions, such as respiratory activity and pupil dilation of subjects in a relaxed state, a stressed-out state, and the state of checking Facebook. The three conditions were very different physically, Facebook lookers having been found to be highly aroused, in a positive way—what the researchers called a "core flow state."

That explains why people might be drawn to check social sites so often, even if it's not the best use of their time. Larry Rosen, Ph.D., the author of *iDisorder: Understanding Our Obsession with Technology and Overcoming Its Hold on Us*, says social media is stimulating, but that too much stimulation can overload our brains and eventually feed into insomnia, anxiety, or depression. "Talking to one friend for two hours is stimulating intellectually, maybe, but calming overall. Two hours spent texting, Skyping, and tweeting to multiple friends overloads us and makes it more difficult for us to enjoy what we want to enjoy." He recommends "resetting" your brain every hour, by turning away from your devices for at least ten to fifteen minutes.

Among the plugged-in masses roam a few Luddites who consciously avoid screens. Jacob, a thirty-eight-year-old midwesterner with a successful career, a love of throwing parties, and many close friends, uses e-mail for work and for keeping in touch with pals, but still doesn't have a cell phone (and therefore never texts), doesn't IM, and isn't on a single social media Web site.

Jacob has been on his fiancée's Facebook page to see photos she'd like to show him, but he doesn't get the appeal. "It's not that I'm actively resisting it; I just don't understand it. It may as well be the *Better Homes and Gardens* Web site; it's just not somewhere I would want to go online. With smartphones or video games or Facebook, all of these things, it's people whittling away their lives through consumer electronics, instead of taking time to think. As for Facebook, it gives me no joy to see so-and-so's kid from the outer rung of my acquaintances. I'm not interested in a lot of weak friendships. I'm interested in stronger ties."

Sunny Updates

Perhaps the scariest dispatch on the fate of friendship in the Internet age was a widely referenced 2006 study concluding that Americans' "core discussion groups"—those people with whom they discuss "important matters"—decreased 28 percent from 1985 to 2004. Subjects who included a friend in this group (as opposed to a spouse or family member) decreased from 73 percent to 51 percent. Many popular articles and books used this as evidence of a decline of real friends amid all our "friending."

But since then, other studies have provided counterevidence to these glum numbers (and one of the 2006 study's own authors has declared doubts about its methodology and conclusion). When Hua Wang and Barry Wellman, Ph.D., compared friendship network size from 2002 to 2007 (and actually asked people directly about their friendships, rather than ask people with whom they discussed important matters, as the 2006 study did), they found that not only did the number of friendships rise, but it rose especially for "heavy" Internet users, who had the most friends both on- and off-line. In 2007, the average American adult had about ten friends with whom they met or spoke at least weekly, with a few additional "virtual" friends and "migratory" friends—online friends who made the leap to real-world interaction.

Keith Hampton and colleagues discovered in 2011 that while core discussion networks are smaller and more family-centered than in the past, social isolation has not at all increased in the last twenty years. In fact, new media use is associated with larger *and* more diverse core networks.

We know that friendship networks aren't shrinking in the digital age, but are the friends we have as close to us as our pre-Internet era ones? It's a complicated and open question, but akin to Toni Bernhard's description of her relationships pre- and post-illness, there is some evidence that online interaction can foster intimacy more than we might intuitively believe. For example, a study of middle-aged female bloggers found that those who disclosed more about themselves online had more (and more satisfying) online friendships. (These women still perceived real-life friendships as more likely to be intimate than exclusively online ones, though.)

If those bloggers were to meet some of their online friends in person and then continue interacting both online and off, perhaps their relationships would be ideal. As researcher Lijun Tang, Ph.D., puts it, the more "social spaces a friendship expands into, the more intimate and rewarding it becomes."

Rather than short-circuit our empathy skills, online friend interactions may strengthen them. "On social media sites, people can learn how to treat people by watching how others treat them," says Larry Rosen, in a positive spin on the exhibitionist aspect of these sites. He conducted research on "virtual empathy" and found that practicing being empathic online helps people carry that behavior over into the off-line world. "Real-world" empathy is still more important to people's feelings of social support, though. "You still need strokes in the real world, but adding in virtual friendships is valuable," Rosen says. "I, for one, am looking forward to getting lots of birthday messages on Facebook from friends who otherwise would not remember that day or get in touch."

Rosen has a positive view of the psychological time warp that social media provide as well. "The fact that someone in his sixties can locate an old friend and then connect with him for a brief period of time and reminisce is a pretty amazing thing," he says, even if the two don't go on to maintain their lost-and-found friendship. Rosen cherishes a burst of online correspondence he had with a few men who played on his Little League team when he was a child. "My dad was the coach, and he was always very hard on me. But getting messages from these guys helped me see a different side of him. They remembered him as funny and caring. I was always so wrapped up in how he treated me; it was nice to learn more about him from another perspective. These chats with people from the past can help reconfigure our memories and enhance our understanding of our own lives."

It's Complicated—Just like Real Life

Online friendship, writes Mark Vernon, "has as many cheerleaders as prophets of doom. In a sense, both are right, for whenever human beings come together, it precipitates loneliness and belonging in equal measure and heightens both."

Pete Beatty, age thirty, currently has 1,474 Twitter followers (he's following 854 himself). He's a big baseball fan (and a fan of all Cleveland-based sports teams); those subjects appear frequently in his tweets. He spends at least an hour a day on the site, and over the years he has gotten together with many people he has "met" through his Twitter feed. At least a dozen of those qualify as "migratory" friends, whom he now sees both on- and off-line. Pete considers others who live far away friends even if he hasn't met them. (One such pal started giving Pete sportswriting assignments, making him an employer/friend/follower.) Some tweeters he meets are different from what he'd expected. "Of course, real people are more complex than a series of one hundred forty characters. When you meet face-to-face, you're getting information through multiple data channels." Describing himself half jokingly as "mildly agoraphobic," Pete likes the comfort of online interaction, even though he does push himself to go out. He also happens to hate talking on the phone, which he views as invasive. By all counts, Pete seems like a model Twitter enthusiast, one who is sharing and learning about his interests, connecting with friends on- and off-line, and even scoring gigs and dates as he exercises his verbal wit.

Yet here is what Pete says about Twitter: "I started my account in the summer of 2008, as punishment for losing a barroom bet. I thought it was idiotic." Surely he's changed his mind four years later? Nope. "Twitter remains stupid. It's a distraction. It's an absurd one-liner contest. I routinely get frustrated by it after two minutes and log out. But then I get back on. I want people to retweet me, to laugh at my comments. I want to be entertained by celebrities and to reach out to people I have an 'intellectual crush' on. That's what keeps me coming back." Pete doesn't consider his Twitter habit a destructive addiction—"I did quit drinking and smoking"—but he tried to curb it once by having a friend change his password, so he could go on a Twitter "cleanse." "After about three days, I caved and begged her for the password. I just really wanted to make a joke about fantasy baseball."

Dissatisfaction with how we're interacting with friends these days is hard to separate out from life stage changes. One friend confided to me, "I feel like when I was younger and single, I used to call certain friends, and we would sit there and talk about our feelings. Often the conversation would be pretty one-sided, like you would call that one friend whose take you liked to tell them about whatever thing was bothering you.

And maybe the next time the bulk of the conversation would be about her and her problem. I feel like I seldom have these kinds of talks with people anymore. If I'm having an existential crisis, the only people who would really listen to me are my parents or my husband. I can't decide if it's because my friends are almost all married now and you just don't call someone who doesn't live alone and go 'blah, blah, blah' or if it's because we just don't talk on the phone that much as a society anymore." This observation is a poignant reminder that even though texting, IMing, and e-mailing let us "touch base" with our closest friends, we sometimes need to pick up that suddenly invasive and clunky phone and settle in for a long talk (or, better yet, a long listen).

All this ambivalence about online friendship reflects competing impulses. Do we like novelty or do we prefer the known and the comfortable? Do we want the attention and status of a lot of Twitter followers and Facebook likes, or do we want to run and hide from public scrutiny? Do we want to hand-select new friends from all over the world, friends who reflect who we are now, or do we want to go back in time and nurture the friends who got us to this point? The possibilities are endless, and choice can terrorize as well as liberate us. We might feel as though these options are being shoved down our throats, but as Pamela Rutledge, the media and psychology researcher, says, "The real question is not 'What is social media doing to us?' Social media *is* us. We created it."

Growing Up Online

Those over thirty can likely divide life into the pre- and post-Internet eras. They made friends before online socializing proliferated, and now they maintain those friends (and sometimes make new ones) online. But what is it like for younger people who have no "before" and "after," whose friends have always existed in person and on screens? Arikia Millikan, now twenty-five, got her first e-mail account when she was eight years old, after her mom got her a Hewlett-Packard personal computer. In high school, she started wandering into online chat rooms. "I was drawn to the kind of disjointed interaction it offered—where you could walk away from the computer and come back and resume the conversation later."

Near the end of Arikia's freshman year of high school in Gainesville, Florida, when she was fourteen, her mother found an e-mail to her from a boy in her class. "It was sexual, but it was jokey—just innocent kid stuff. But my mom completely freaked out. She ended our Internet subscription. So all through high school I had to walk to the public library to get online." She could IM there at the library but felt very distant from her peers who had constant access. "There was this whole conversation I was missing out on, and relationships I couldn't forge. Knowing that I was missing out probably drove the tech obsession I later developed."

Before going off to college at the University of Michigan in 2004, Arikia got a new laptop. It happened to be the year that Facebook first became accessible to colleges other than Harvard. "You would meet someone in a class or something, and then you would immediately look them up on Facebook," she says. "You would have way more information about that person than was ever possible before."

Reading Facebook profiles entailed more than just checking out someone's favorite bands or movies, Arikia says. It was an intuitive process that yielded an overall impression of someone.

"Throughout college I became really good friends with people who were really different from me, opposite in their political views, for instance. Facebook just framed the conversation going forward. You had access to things that person hadn't told you, but that were fair game information to discuss."

I wondered if maintaining her own Facebook page was a stressful game of image maintenance, given how crucial these profiles were to social life. "I was always pretty authentic," she says. "But you want to have your best face forward, so there's the process of deleting unflattering pictures and crafting your updates to reflect the best parts of your personality. I was probably less self-conscious than other people about photos that were potentially incriminating, like of me drinking at a party." Students were warned by administrators, in fact, not to post comments or photos that they wouldn't want a future employer to see. "I was very quick to take the position that if a future employer was going to hold something silly I did in college against me, that wasn't the kind of employer that I'd want to work for." Spoken like a stereotypical millennial!

Yet it was a prophetic notion: Arikia is now an online editor at *Wired*, the tech and science magazine. Her natural love for online socializing turned into a job offer when a *Wired* editor started following her tweets and gave her a few freelance projects to work on. Still close to many of her college friends, she believes she has personally influenced several of them to move to New York, where she headed right after graduation. "I think some of my friends were drawn to come here, based on my portrayal of my experiences in the city on social media.

"I'm always online," Arikia says. "I never disconnect, except when I sleep. I probably go to about four events a week; most are media or science related; it's an opportunity for people to get together and see friends from the Internet and meet new people." In an ironic twist, Arikia met her roommate, whom she considers her best friend, the old-fashioned way, at a bar. But their first conversation was about none other than Facebook. "We were thrilled to find another person who understood social media as much as the other.

"Social media has made such a big difference in my well-being that I like to show other people that it can be a really enjoyable part of life," Arikia says. "For me it's really been the vessel to solidify friendships that I can't imagine would have formed, or would have formed so quickly, if it wasn't for the availability of the communication media."

As for those who say people of her generation are empathyless narcissists without real friends, Arikia says, "Anyone who would say that has obviously not experienced the full benefits of the Internet or even given it a chance. I feel sorry for them."

Even if Arikia's social habits are extreme, she's the canary in the coal mine for today's teenagers, who may have e-mailed since kindergarten. But the bad news is that those who haven't fully developed their social skills do seem to suffer from a lack of "face time" with their peers (and their parents). Researchers from Stanford University surveyed preteen girls and found that the more time these young women spent in front of screens, the more likely they were to show signs of poor emotional and social health, such as low self-confidence and having more friends whom their parents regarded as bad influences on them. Those who reported more "in-person" chats with friends, parents, or siblings, however, showed opposite health outcomes. On average, the girls used electronic media for a startling 6.9 hours per day, and talked face-to-face with people for a total of 2.1 hours per day.

Larry Rosen has studied the online life of American teens and has found some pretty jarring adverse effects, although, again, these effects correlate with very high usage rates. Though social media sites can coax timid kids out of their shells, heavy users are more prone to anxiety, insomnia, depression, and stomachaches. Adolescent Facebook fans are more narcissistic and even more aggressive than their peers who use social media lightly or not at all.

Still, Rosen is not disturbed by his results because he thinks teens are tempering their use, making themselves less vulnerable to these extreme consequences to their health and development. "Teenagers have taken social communications, texting, IMing, all of those things together, as their world. It's the focal point for them," Rosen says. "But the pendulum is starting to swing back, as teens are realizing that the virtual world is not enough—that you need real-life contact, too. They were the first to embrace this world, so they will be the first to experience a rebound effect."

For average teens using social media in moderation, online behavior probably does not cause any problems they wouldn't have without their devices. Danah Boyd, Ph.D., a fellow at the Berkman Center for Internet and Society at Harvard, has concluded that online adolescent life—surprise, surprise—mostly mirrors the kids' off-line lives. Because children are not allowed to explore the real world as freely as earlier generations of kids did, it's only natural for them to express their curiosity by running around the corners of the Internet, she argues. Teens from troubled families are more at risk for getting into dangerous behaviors online than teens from solid, supportive homes. And finally, online bullying is no more rampant than off-line bullying, she insists. While kids might be mean on Facebook or keep secrets from their parents tucked away behind passwords, those are timeless teen behaviors, not a cause for undue worry or alarm. (However, as we've seen with recent highly publicized teen suicides, online attacks can proliferate via social media and leave a vulnerable young person feeling overwhelmed, overexposed, and publicly humiliated. So even though bullying is not necessarily more common online, it does have the potential to be much more damaging than the back-of-the-bus harassment of the off-line teen universe.)

As for very young children who now play with Daddy's iPad and Skype with Grandma, it remains to be seen how their social skills will be affected by their screen time. Once again, parents should encourage moderation in social media use. In other words, more playdates and fewer hours with the electronic babysitter.

Friends Make the (Online) World Go Round

We may not yet know precisely how our newly "pervasive and persistent" friendships will shape us going forward. But just because some of our friends might now be online-only doesn't mean they won't continue to steer our thoughts and behaviors. In *Connected,* James Fowler and Nicholas Christakis conclude that "the eating, drinking, and smoking habits of our friends who live hundreds of miles away appear to have as much influence as the habits of our friends who live next door. This means that ideas about behavior can spread even in the absence of frequent direct personal contact." They also predict that deeper connections will still affect our attitudes and actions more than weaker ones—even if both of those are displayed the same way in our social media feeds. (One preference that apparently doesn't spread through online social networks is musical tastes. With the exception of classical and jazz, genres did not go viral among a group of college students whose Facebook activity was monitored for four years. This could be because tastes are solidified in adolescence or because only close friends cajole us to listen to their favorite tunes while online acquaintances don't sway us away from our beloved playlists.)

Your friends are increasingly becoming news editors—telling you, via their posts and tweets, which articles to read amid the content blizzard and influencing your view of a world that, as Clay Shirky wrote, used to be shown to you by a small group of professional journalists. And friends are increasingly becoming your personal shopping advisors, too, so that they now have a power that marketers are desperately trying to harness. Irfan Kamal, the senior vice president of digital and social strategy for the advertising firm Ogilvy & Mather, recently referred to internal research showing that people were two to seven times more likely to go to fast-food restaurants when they were exposed to them via their friends' social media pages.

Paul Adams, a researcher and designer at Facebook, has said that the Web is rebuilding itself around social groups (as opposed to page links). "People turn to their friends to make decisions," he declares in a presentation he gave. Influence, he adds, spreads through regular people like you and me, not special "influencers," in the marketing parlance of the last decade. You are the connector between your independent groups of friends, and it's the people closest to you, not charismatic types in your outer circles, who are most influential.

Some of us are more easily influenced than others, Adams points out. If you are risk averse or have deep-seated habits, you might not be swayed by your friend's enthusiastic review of a Costa Rican adventure tour. But that friend's word is still more influential than that of a celebrity, no matter how famous. "Celebrities make us aware of products," Adams says, "but they don't change behavior." It's our friends who are guiding us through the morass of data coming our way each day. "We live in a world of exponentially increasing information. It's not going to stop. We can't absorb it all, so we're increasing our reliance on our social networks to make decisions."

Friendfluence can also change the political landscape more easily than ever, also courtesy of social media. Tina Rosenberg, a writer and expert on peer pressure, looked into the effect of the site Friendfactor.org on the recent gay marriage legislation battle in New York State. Instead of a Twitter or Facebook campaign that blasts all contacts with messages at once, about a thousand people created pages on Friendfactor, which sent personal messages to their close friends. Each friend was asked to make a call to his or her representative about supporting gay rights. After each did so, the asker of the favor got a message saying his friend had complied.

"High-risk activism is a 'strong-ties' phenomenon," Rosenberg writes. "The Egyptians in Tahrir Square were motivated by their connections to others in the group around them. They wanted to go out in the street and be daring and take risks with their friends. Real friends can flip a switch that turns ordinary people into heroes. Online 'friends' cannot." The Friendfactor.org campaign generated about eight thousand calls to state senators. The legislation indeed passed.

YOU'VE SURELY EXPERIENCED THE BIZARRE modern phenomenon of being in a crowded public setting, such as a café, where nearly all the patrons are gazing at a smartphone or laptop. The dimension of flesh-and-blood folks is a mere background set of the social plane within which those people are actually engaging. The newest social media sites are attempting to crash through and unite those two spheres. At a recent tech conference, no fewer than eleven apps were being peddled as tools for alerting people to locate potential friends in their physical midst, on the basis of common interests or social connections.

The thought of a fellow, say, Stevie Wonder fan tapping me on the shoulder while I'm having a coffee break does not appeal to me at all. Nonetheless millions of dollars are being poured into these apps at the moment, meaning they must excite some. Michelle Norgan, a cofounder of one such app, Kismet, was quoted in a *New York Times* article as saying, "It is a push to get people to get their heads out of their phones and back into that one-on-one." Yet again, the basic impulse behind online friendship is revealed: With all of our gadgets and platforms we're only trying to create a friendly Shangri-la, where friends from past, present, and future commingle freely right beside us, space and time be damned.

CHAPTER 8

Making the Most of Friendfluence

RENEE YOUNG MET HER BEST FRIEND, Connie, fifteen years ago when Connie was recommended to her as a troubled-cat specialist. Renee had purchased an out-of-control kitten from the pet shop and had asked the humane society for help. "I really didn't want to call Connie," Renee recalls. "I thought she would be some crazy lady with 25 cats who was going to tell me what a horrible person I was for buying an animal from a pet store. To my surprise, a lady with a lovely Southern accent answered the phone. She was just darling." Connie, who is twenty-five years older than Renee, soon thereafter charmed Renee's children by getting down on her hands and knees to train their kitten while clad in an elegant yellow silk suit. Now Connie and Renee speak every day and have dinner together each Friday night.

"From Connie I've learned how to give people the benefit of the doubt, not judge people, to truly be kind to animals, identify valuable antiques, and make amazing mashed potatoes," Renee writes. "Along the way we have buried family and friends, weathered two divorces (mine and her daughter's), lost jobs, found new jobs, seen my children get through childhood and go on to college, seen her become a grandmother, seen boyfriends come and go (mine), lived through one heart attack (hers), rescued countless animals, gained and lost many pounds, and had so many adventures I cannot remember them all. I cannot imagine my life without my beloved Connie."

We're built to create and crave friendship, and friends can shape us just as much as, and sometimes more than, our families. Our parents (save the truly dysfunctional ones) love us naturally; we often have to earn and elicit a potential friend's trust and affection. Our parents must keep us in the family unit; we must insert ourselves into peer groups that might not initially hold out a seat and could shut us out at any time. A sibling has to share space and time with us; a potential friend could choose other playmates if she wished to.

The child's task of learning to make and get along with friends is crucial: It prepares her for getting along with people in all contexts—future coworkers and bosses, future romantic partners, roommates, neighbors, acquaintances, and, of course, future friends. Social skills form the foundation of many of the endeavors and circumstances that bring us happiness, success, and good health. As John Cacioppo, Ph.D., director of the Center for Cognitive and Social Neuroscience at the University of Chicago, points out, humans are "obligatory gregarious," simply meaning we need to be with and to like other people. We have to feel a sense of belonging from people outside of our families. Whether or not we get that sense is the highest-stakes example of friendfluence: To have or not have friends at any stage in life influences everything from your personality and outlook on the world to your ability to fight off disease.

Beyond the basic hunger for connection that minimal friend forging satisfies, your mix and intensity of friendships influence you in myriad ways, across different areas of your life. I'm guessing you don't often think of those effects carefully or holistically, though. It's acceptable to pick apart a romantic relationship for hours on end, it's normal to debate the merits of different parenting theories—with gusto or even self-righteousness—but considering one's friendships overall (e.g., apart from a specific friend crisis or annoyance) is not generally done. Maybe that's because friendships are seen as a respite from these other, more challenging relationships. Or maybe it's because we simply don't realize that friends are worthy of more scrutiny. The influence of any one friend might spread thinner than that of any one mother, father, spouse, or sibling, but the influence of all of your friends spreads thick.

We seek out dating advice, but maybe the best way to find a true love is to get in with a group of friends who share our interests and will invite us to places where we'll be more likely to find romantic prospects. Furthermore, those friends will give us companionship and advice through the travails of dating, making us more secure and confident—the very qualities that attract most available men and women. We seek out marital advice on dealing with a spouse, but couples who cultivate independent friendships and mutual friends are stronger than those who "cocoon." We seek out parenting advice, but a support system lessens the stress of caring for little ones and exposes children to a variety of adult role models and perspectives, giving the proverbial "it takes a village" a chance to be more than just political rhetoric. We seek out health and weight-loss advice, but the most effective plan might be to hang out with fit friends. Not only do they make it easier for us to eat better and work out by setting an example and dragging us along to active outings, but they also provide the human connection that fosters robust physiological characteristics, such as lower blood pressure and increased immunity. We seek out career and résumé-building advice, but friendships at work boost productivity and opportunities. Plus, spending time informally with friends who share our same vocational interests keeps us in the know in terms of available jobs, current ways of thinking in a particular industry, and general gossip that we can use to our advantage. We seek out self-improvement, but what we really need is a little help from our friends.

Yet regarding friends as the way to get what you want out of life can sound creepily opportunistic. Of course, friendship wouldn't be friendship if it were founded primarily on selfish expectations: It's a two-way dynamic, and you can't simply shop for friends who will fulfill your needs and help you realize your dreams. After all, you have to experience a visceral type of attraction to a potential friend. Then both of you have to ramp up your levels of closeness and intimacy over time. You have to be a friend, with all the work that entails, to have a friend.

Mark Vernon worries that thinking about the benefits of buddies obscures the bigger picture: "By placing you yourself at the center of the universe, as self-help almost invariably does, it treats everyone else in the universe as bit players in the story of your life. Hence friends cease to be other people, whom you might know and love as persons in their own right, and are regarded as sources for the various elements that you need in your life—one friend to shop with, another friend to cry with, another again to laugh with, and someone else to rebel with. Friends, in short, as service providers. And as everyone knows, the minute your friends start to feel used, for all that they may otherwise be happy to be useful, is the minute your friendship starts to fall apart." Self-help focuses on a cost-benefit analysis, Vernon writes, where people are told to smile or help strangers "not because it is good to be grateful and friendly but because exhibiting gratitude and friendliness comes with the promise of personal happiness in return. . . . It's the morality not of do as you would be done unto, but do because it delivers."

I don't think you can separate morality from pleasure in the case of a functional friendship, one that has the potential to encourage the best behavior we're capable of while also making us happy—precisely as nature would have it, given friendship's potential to help us meet our most basic needs. You can nurture real friendships without a thought to reciprocity and extend yourself for others because it is morally correct, *and* you can also choose your friends more carefully and modulate the intensity of your existing friendships. To have friends whose values and lifestyles align with your aspirations is a very worthy long-term goal. If you consciously and consistently treasure good friends and phase out or distance yourself from those who have a negative influence on you, eventually your genuine friendships, based on intangibles such as love and caring, will provide the tangible benefits that the science of friendship showcases.

You don't love dear friends for what they can give you; you love them for who they are, and in fact you get more pleasure out of being a good influence on them than you'll ever get from the perks they provide. But their good influence on you is a worthy side effect. In fact, the best kind of friend will be, above all else, a positive moral influence; he or she will make you a better person.

Friendship's greatest thrill is knowing someone, a unique human being, very well, sharing experiences with him, and having the power to make his life easier and better. Happiness is the by-product of that kind of commitment; it's not something you can directly chase and catch. You don't begin a friendship because you are anticipating the gratitude and contentment that Renee feels when she looks back on her fifteen years of friendship with Connie, the crazy cat lady turned soul mate. But holding in your line of sight all the benefits that good friends can bring is arguably the best self-help plan you could possibly enact.

STEPPING BACK AND SIZING UP YOUR FRIENDSHIPS

Since May 2007, psychologist Irene Levine, Ph.D., author of *Best Friends Forever: Surviving a Breakup with Your Best Friend*, has been dispensing advice on the friend quandaries that pour into her Web site, the Friendship Blog (she's the Friendship Doctor on *Psychology Today*'s Web site). "I've become more acutely aware of people's different friendship styles," she says. "I've realized that I'm probably more introverted and I've come to accept that I may have less of a need for a huge number of friendships in my life than others. The friendships I like are dyadic—close, intimate friends. I hate cocktail parties, I dread group lunches." Her advice for those who, like herself, draw energy from being alone and engaging one-on-one rather than from the whirl of a big gathering? "Know your own style, but sometimes go beyond your innate nature and consciously make an effort to get more comfortable in group settings. I make myself go to luncheons sometimes, and I feel better after I do it."

Though one friend is vastly preferable to no friends, says John Cacioppo, who is also the author of *Loneliness: Human Nature and the Need for Social Connection*, there is no simple answer to the question of how many friends are enough. "Each of us inherits from our parents a certain level of need for social inclusion (also expressed as sensitivity to the pain of social exclusion)," he writes. "This individual . . . propensity operates like a thermostat, turning on and off distress signals depending on whether or not our individual need for connection is being met." Some are devastated by social slights; others barely notice them. Some shriek in anticipation of seeing all their friends at one big celebratory event; others cringe at the idea and would much rather have a cup of tea with each pal individually.

Inborn temperament aside, satisfaction with your friends depends on the ones you have. "If you have one friend," Cacioppo says, "and they don't really provide the support you need across all of your life domains, then you'll feel like you're missing something." That's why people with kids become friends with other parents, he says; it's in part because those are the people they are running into, but also because "those are the people who have the most important information about that domain."

If you've developed a baseball obsession and have no one with whom you can share your postgame reactions, you can feel lonely, even if you have many friends with whom to talk about your job or love life. The particular problems you are facing can also be a metric for judging whether you have enough friends. "What others do is to help you overcome life's challenges," Cacioppo says. "How many life challenges do you have and what is the nature of the challenges that you're facing?" Parents may gravitate toward other parents, for instance, but moms and dads of kids with disabilities might need another parent of a special-needs child for a companion, someone who will truly understand their situation.

Having a variety of friends, each of whom relates to either your current lifestyle or one of your interests or aspirations, is ideal. Another, rather fun way to analyze your social life is to categorize your friends by personality style. You might find that at this point in your life you could really benefit from having a friend with a style that is currently missing among yours. Tom Rath, of the Gallup Organization, identifies eight "vital roles" in his book, based on extensive interviews with people about what they get from their friends.

"Builders" are motivators who push you to succeed. "Champions" stand up for you and have your back—even when you're not around. "Collaborators" share your interests or beliefs. "Companions" are blood-brothers who will sacrifice for the sake of the friendship. "Connectors" introduce you to others and invite you to events and parties. "Energizers" are fun, positive friends who know how to cheer you up when you're sad and calm you down when you're rattled. "Mind Openers" expose you to new ideas, thoughtfully question your opinions, and help you innovate. Finally, "Navigators" help you weigh decisions and map out realistic paths to your goals and dreams. (It's also illuminating to categorize yourself on the basis of these roles. You are likely a bit different with each friend, but yours could collectively point out some consistencies in what you give to them.)

Given what we know about the value of "unlikely" friendships, I think it's also worth looking around your birthday dinner table to see if everyone in attendance is just like you demographically. If so, make an effort to get closer to acquaintances who check off different items on the census than you do, or boldly strike up conversations with people outside your circle, in the hope that you will develop a friendship with someone from "the other America," whatever that is to you.

Then again, feeling unsatisfied with your birthday guests could have nothing to do with them, but rather be a matter of your idealized vision of what a group of friends should be. For one thing, your pals are probably not all close to one another. The stable, tight, exclusive groups so often portrayed on TV and in films, from *Friends* to *Friends with Kids*, are rare. In real life, groups turn over as individual members literally move away or metaphorically head in different directions. School and work friends don't always magically mix, nor will your gang and a spouse's necessarily gel into a new cohesive supergroup. If you're lucky enough to be a part of an enduring clan of friends, be grateful for the closeness, security, and boon of shared memories. But don't be afraid to wander a bit either, to see if the role your tribe has cast for you is indeed the one you'd like to keep playing full-time.

Your friend analysis should leave you resolved to strengthen your relationships with the people who have the best influence on you. It could also leave you convinced that you should gently taper off the company of certain others. You might even consider, in dramatic cases, breaking up with a friend who is especially difficult.

In her book, Levine offers some questions to ask yourself when you're determining if a friend is worth keeping or should be cut loose. For example: Does scheduling time together feel like an obligation rather than a pleasure? Is the friendship a constant source of irritation? Do one or both of you show habitually bad judgment? Do you feel emotionally drained when you are together? Finally, can you trust each other to keep confidences?

Winding down a friendship with someone who is needy or dependent is especially tough because you might have an ethical obligation to look out for her if you've been friends for a while. "It's very difficult to draw a line in those situations," says psychologist Terri Apter. She nursed you through a bad time years ago, perhaps, but now she's calling in the middle of the night, disrupting your family life. "Sometimes you'll see that a friend is really going downhill; perhaps she's an alcoholic, or she's depressed. What can you do in order to save her? There is no clear answer. You have to say to yourself, 'How should I think about this? How can I balance what is in my life now? What can I do for her and what can't I do for her? And what do I owe her versus what I owe my partner, my kids, my clients, or myself?' "

Is there an optimal way to cut off a friend? As with so much in managing friendships, it depends. Andrea Bonior suggests "slowly backing away" as a first-line strategy. "You can respond to their calls not quite as quickly, respectfully decline invites, and not ask as many questions when together," she says. "This can really only work if the other person is willing to take those cues and back away herself, though." If, on the other hand, the friend confronts the one who initiated the breakup with this behavior, the latter will often shy away from an outright conflict and will make up excuses such as "Oh, I've been busy at work." Then the initiator is right back where she started.

Truly dysfunctional relationships, such as when one person has a borderline or narcissistic personality disorder, require a confrontation that is practically destined to get ugly. In these cases, Bonior says, a direct breakup declaration is usually necessary: "The other person is not going to respond well, and a lot of times it's going to end up in an explosion."

Technological advances have made friend breakups even trickier, Bonior says. "E-mail and social media don't allow some friendships to die a natural death. It used to be that if a friendship was subpar, all it would take would be a switch in circumstance—somebody moved or had kids—and the friendship would dissolve. Now, that's not an excuse. It's a lot harder to back away from a toxic or even an 'eh' relationship. You have to take a stance, by 'unfriending' someone." What's more, people who fall out with disturbed friends might have to do extra online postbreakup "cleanups"—Bonior has heard of betrayed friends smearing one another online. What used to be a vituperative letter or phone call could now be a hard-to-erase mark on your cyberspace profile.

Friend breakups can be excruciatingly painful, especially since as discussed in Chapter 6, there's no standard protocol and no sanctioned grieving process for the aftermath. (Taking a day off work after the demise of a long-term romantic relationship would be totally acceptable in some offices. Asking for the same after a falling-out with a friend would be met with skepticism or even mocking incredulity.)

UNDERSTANDING AND CURBING LONELINESS

Of all those letters on friend dilemmas that Irene Levine has received from her readers (who are primarily female), perhaps the most common theme is loneliness. "There is a large group of women who feel friendless and are completely baffled by it," she says. "They're looking for ways to connect, and if they haven't had success by the time they've reached an older age, they are less confident in their abilities to make friends. It starts to feel like a fatal flaw in a culture where women are valued more if they are more popular."

Cacioppo's research makes the case that loneliness is nature's painful cue for us to connect with others. Unfortunately, though, its effects over time are so pervasive and damaging to body and mind that they can prevent us from being able to make the very friends who would relieve the primal ache.

The ability to deal with people—the basic skill of friendship—is largely what made humans what we are, Cacioppo argues. "Most neuroscientists now agree that, over a period of tens of thousands of years, it was the need to send and receive, interpret and relay increasingly complex social cues that drove the expansion of, and greater interconnectedness within, the cortical mantle of the human brain."

Chronic loneliness, Cacioppo has shown, is associated with depression, cardiovascular problems, sleep dysfunction, high blood pressure, and an increased risk of dementia in older age. It also triggers thoughts and behaviors that can cause a tragic self-fulfilling prophecy over time: Cacioppo found, for instance, that the lonelier young adults are, the more they withdraw from the world when stressed out. Lonely people who do bother to reach out find positive social interactions less mood boosting than do nonlonely people. When people they know experience a blessing in life, lonely people are less able to feel vicarious happiness for them. This isolates them from the friends they do have: "Studies show that truly enjoying those positives and making the most of them is even more important to the health of a marriage or other intimate relationship than being supportive during hard times," he writes. Lonely people develop a more distrustful and negative view of people in general. "The cynical worldview induced by loneliness, which consists of alienation and little faith in others, in turn has been shown to contribute to actual social rejection. . . . If you maintain a subjective sense of rejection long enough, over time you are far more likely to confront the actual social rejection that you dread."

In another example of how the benefits of feeling connected garner more benefits, while friendlessness begets more and more problems over time, happier and less lonely people make more money than do lonelier souls. Relatedly, a study by Sigal Barsade, Ph.D., and colleagues looked at 650 workers and found that loneliness on the job leads to less productivity. Individuals can feel left out of the office cliques, of course, and sometimes a company's very culture seems to foster friendlessness, via distrust and discouragement of casual socializing among employees.

Though people with severe social difficulties might be especially lonely, those who are lonely are generally just as able to make friends as anyone else. "Feeling lonely does not mean that we have deficient social skills. Problems arise when feeling lonely makes us less likely to employ the skills we have," Cacioppo writes.

Introverts get pegged as awkward or shy, but their social skills are often on par with their more outgoing counterparts. And since they require fewer friends to feel connected, introverts are not lonelier than extraverts. However, once they do feel lonely, it's harder for introverts to get reconnected, Cacioppo says, because they are not out meeting people all the time the way extraverts are.

In work with Christakis and Fowler, Cacioppo has also shown how loneliness spreads through social networks. If a person said he was lonely, two years later his closest connections were 52 percent more likely to report feeling lonely, too. "There is a mechanism by which this happens," Cacioppo says. "Let's say that you and I are neighbors and we're good friends. For some reason unbeknownst to you, I get to feeling lonely. That means I might start acting in a more negative, more self-protective, less generous fashion than you. Over time, you might cease to be my friend, because it's just better to not deal with me. I became lonely and I transmitted that to you, because look, you lost a friend. Even though we haven't moved physically, we've moved closer to the edge of the social network because we have fewer friends now."

This mechanism is turbocharged by the fact that negative social interactions weigh much more heavily on people than positive ones. That is, you could have one day of completely pleasant exchanges with coworkers, cashiers, fellow commuters, and so forth but then get put in a hard-to-shake bad mood after one tiff with a crabby waitress. What this means, says Cacioppo, is that "if we were living next to each other and we were friends, and I'm interacting with you in a more negative fashion, you are now more likely to interact with somebody else in a negative fashion. You're not only losing me as a friend, you might compromise your other friendships as well. And those friends might turn around and have more negative interactions."

Cacioppo and colleagues have made a very powerful case for the dangers of friendlessness to one's mental outlook, happiness, and health and to the functioning of communities and our society as a whole. But his ultimate message is an empowering one: "It is my belief that, with a little encouragement, most anyone can emerge from the prison of distorted social cognition and learn to modify self-defeating interactions. What feels like solitary confinement, in other words, need not be a life sentence."

"People tell me all the time that they feel like stalkers," says Andrea Bonior of her clients who want to strike up more friendships. If you have a conversation with someone and feel a friendly spark, should you ask him or her to join you for coffee? Or is that just weird? "With dating, people know the script," Bonior says. "You know whether or not someone is interested, after a short time. Honestly, it's really scary to say, 'I'm going to go out and meet friends,' because it's such an ambiguous process."

That is precisely what the journalist and Web producer Rachel Bertsche did, however: She spent a year going on fifty-two "friend dates" to combat the loneliness she felt after moving to Chicago for her husband's job. Though Bertsche is happily married and gets a lot of support from long-distance family and friends, she missed in-person contact with girlfriends and decided to launch a challenging experiment to regain the level of connectedness she had had before moving.

Bertsche, who chronicled her friend-dating adventures in *MWF Seeking BFF: My Yearlong Search for a New Best Friend*, says that though she feared courting people for friendship, most were happy to be asked on friend dates. "There is a stigma attached to loneliness," she says. "If you say you don't have friends, people might wonder what is wrong with you. I was afraid to tell people I was looking for friends because I didn't want to look like a loser. But actually, a lot of people can relate."

An unexpected result of Bertsche's experiment (which did yield her many close friendships) was a sense of personal expansion. "Before, I was not someone who did things outside of my comfort zone. So a lot of things I did during my year of finding friends, such as signing up for a book club, a cooking class, and an improvisational comedy class, helped me branch out. Before, I would never have signed up for a class unless I was taking it with someone I already knew. The improv class was not only a satisfying challenge, it was also helpful in connecting with people because those first conversations involve some improvisation. I talk to everyone in my path now. I've built up that social muscle."

Bertsche's need for friends was brought on by a relocation, but sometimes placing yourself in a new environment is a good way to improve your social life, even if you don't actively put yourself "out there." In his book, Cacioppo recalls Jane Jacobs's 1960s era defense of small, vital communities where people live and work close by. "She writes about the greater trust and sense of connections as well as the enriching, serendipitous encounters that result. I can attest to her insight, because my wife and I live in just such an urban village . . . where neighbors know one another's children and pets and keep up with the progress of one another's plantings beside the doorsteps."

The culture of a particular place can impact individuals and their friendships. "If you live in a neighborhood that is highly materialistic and competitive," Cacioppo says, "that could make negative social interactions carry a little more weight than they would if people were looking out for each other and were more generous because they weren't trying to increase their status by decreasing someone else's." Or for example, how scared people are of crime in a neighborhood is only partly a function of how much crime there actually is. It's also related to how many crime-related shows people watch, he says. "How much crime TV people watch is one of the neighborhood predictors of loneliness. People feel isolated because they think it's unsafe to go out."

Joseph, forty-four, recalls two times when a fresh locale stimulated his social life. "I was very isolated when I was in my early twenties," he says. "One summer I took a job on a Greek island. The people there were either happy, friendly vacationers or employees in the restaurants and other touristy places. You'd see the same faces all the time, and since many people were from other countries, there was an instant camaraderie. It made it very easy to make friends and casually hang out on the beach or at a bar." More recently, Joseph relocated from a bustling Manhattan neighborhood to a residential area in Queens. When he visited his neighborhood Starbucks back in Manhattan, Joseph could never tell who among the other patrons were actual neighbors, since some were tourists and others worked in the area but had no desire to make pals there. Now he regularly holds court at his local Starbucks, waving to half the patrons and chatting with others. Since there are no large office buildings or tourist attractions nearby, a small-town feeling presides. (The post office and elementary school are in fact right across the street from the café.) "There is more social interaction and more pleasantries are exchanged here," he says. "It's like there is more permeability between people. That opens the door to friendship."

Incidentally, even if you can't change your circumstances at the moment and aren't having success drumming up friends through conscious efforts, you can assuage loneliness with a nostalgia session. One study found that lonely people who indulged in memories of the past did so to elevate their feelings of social support. Combing through old pictures and letters from friends for a bit (you don't want to become too embroiled in the past, at the expense of building a future for yourself) can remind you that you had warm connections before and therefore have it within you to cultivate them once again.

NEARLY EVERYONE FEELS SOME ANXIETY when interacting with strangers or attempting to create relationships of any kind, but some people who truly struggle suffer from social anxiety disorder. Obviously this makes it harder for them to make friends, though it doesn't have any bearing on their ability to be a wonderful friend to those they know. Some symptoms of social anxiety disorder are feeling extremely self-conscious in front of others, experiencing such physical changes as blushing and sweating, and worrying for days or weeks about an upcoming event.

The technique that psychologists commonly employ to help social anxiety sufferers, called cognitive behavioral therapy, can squash subclinical symptoms of the condition, too. "The therapist will give patients little homework assignments," says Bonior. "It might sound silly, but for people who are not used to face-to-face interactions, an example would be making small talk with the grocery store clerk, practicing eye contact while speaking, or asking a follow-up question to someone during a conversation. These exercises get the social juices flowing." Small victories engender a greater willingness to experiment. The overwhelming buzz of the social world quiets down.

For those with autism, meeting people is not only anxiety provoking but also baffling. As a high-functioning woman with Asperger's syndrome, Lynne Soraya is in a preferable position to many on the spectrum, given her high intelligence and self-awareness about her own limitations. Still, the comparison doesn't erase the many times she has felt lonely and cursed with a strong desire to make friends yet a weak ability to do so effectively. Now in her thirties, Lynne lives in the Midwest with her husband and is a stepmother to three sons. "I do feel lonely at times, but I feel like much more of a social success than I've ever been," she writes. Lynne conceptualizes interacting with those not on the autism spectrum as "speaking across a neurological barrier," akin to negotiating cross-cultural relationships. "Once you understand the differences in thought, approach, and communication, it does a lot to bridge the gap. That said, I still feel misjudged at times."

As a child, Lynne often had at least one confidante, but when those relationships inevitably ended, she took it extremely hard. "I believed something was terribly wrong with me and that people didn't like me." Reflecting back, she realizes she didn't express caring in typical ways, leaving people with the impression that she didn't feel affection for them. When kids did reach out to her, she often missed the overtures. "I'd been rejected and bullied so often, I learned to be skeptical of whether people really liked me or not, so I'd keep myself at a distance."

These days, Lynne tends to make friends in structured settings where others share her interests, such as on a volunteer project. The thought of befriending a complete stranger still makes her nervous. And giving her friends what they want has yet to come naturally. "I might be very motivated to try and fix a problem for someone, but I might miss when they need an emotionally comforting gesture. It's the Golden Rule: I do unto others as I would have them do unto me. But, because they are different than me, what I would want is not necessarily what they would want."

Still, Lynne has actively improved her body language and

people-reading skills; she credits avid observation of others, psychology books, and a high school theater class for helping her make great strides. You don't need to change who you are to find a true friend. In fact, such an endeavor would defeat the purpose of having a pal who really gets you and relates to you, but fitting in is often essential to group social life—the garden from which those true friends are plucked.

COMMON FRIEND PREDICAMENTS AND HOW TO BE A GOOD FRIEND

Jacob, the Luddite from Chapter 7, who eschews Facebook and doesn't have a cell phone, is an exemplary friend, with many long-term, loyal ones to whom he is equally devoted. "In my twenties and thirties, large gaps between romantic relationships allowed me more time for my friends," he offers as one explanation. "And having a family that is in no way dysfunctional, at least not scandalously so, but that doesn't give a lot of sustenance made me more interested in close friendship." Jacob suspects he learned how to be a good friend in college, from two guys he met there. "It becomes a virtuous circle. When you realize the joy you can get out of them, you invest more. Because I've had primarily good experiences, I realize it's worth it to make friendship one of the most important things in life.

"I read a quote once where a famous writer said to another, 'If you're not willing to be bored sometimes, you can't have friends,'" Jacob says. "Sometimes friends are going to drone on about their mother or something that you don't quite care about. But it's not just about what they can do for you, it's a deeper thing. You can't expect to always be entertained, or to always feel like everything is one hundred percent reciprocal." Jacob, who likes to entertain, says, "I'm willing to invite someone to dinner ten times and never see their house, because if you get into the cycle of pettiness, you won't end up having any friends."

Known for taking an active interest in his friends' major decisions and offering them feedback that, if valuable, can sometimes be hard to hear, Jacob admits he doesn't always like it when the tables are turned. "I'm a hypocrite just like everyone else. Part of my desire to have impeccable behavior with friends," he says jokingly, "is my lack of joy at having my less-than-impeccable behavior pointed out to me."

Henry Alford (the etiquette expert introduced in Chapter 7) considers the tricky question of when brutal honesty with close friends morphs into rudeness. "I loved it when a friend whose shoulder I cried on asked me: 'Do you want advice, or do you want support?' " he says. "It's a crucial distinction. If we don't ask this question, then we are forced to rely on intuition and past experience. Is this the kind of person who can handle candor, or is this the type of person who needs the kid gloves? If it's someone you know pretty well, then you'll be able to answer that question. If not, then it's better to err on the side of discretion, unless the stakes are very high (i.e., your friend is about to appear on television and has spinach on her teeth)."

Alford suggests considering the realm (career, family, fashion choices) of an intended critique and whether it is central to that person's identity: "Which of these arenas is she sensitive about? Each of us has his own areas of vulnerability. You can talk dirt about my appearance, but don't touch my intelligence! Usually, if the person makes jokes about a certain aspect of her life, then she's implicitly telling you that it's okay for you to make jokes about it, too."

Daniel Hruschka reviewed studies on the causes of conflict in friendship and found that the most common friendship fights boil down to time commitments. Spending time with someone is a sure indicator that you value him; no one likes to feel undervalued. But there seems to be a paradox here when it comes to friendship: Friends can help us so much in so many areas of life, and we often want to spend more time with them. Yet keeping in touch and planning get-togethers are often relegated to the bottom of the to-do list.

"Because close friends are seen as voluntary and pleasurable, they are the first thing that goes out when domestic responsibilities meet work responsibilities," says sociologist Dalton Conley. "We spend more time with our kids than our parents did, we spend more time working than our parents did, so what else have we given up besides sleep? We've given up friendship. We might be popping Xanax on the go, but we've given up the social cocktail hour."

Now that Jacob is planning to get married, he sees time as a potential enemy to his desire (and hers) to not completely cocoon with his new bride. "I'm working a lot, and if we want a few nights with each other and one for ourselves, and one to just sit there and do nothing, that leaves one night a week to go out with friends. There's no lack of desire; it's just the reality of modern life. If you categorize friends by frequency, there's the 'once a week,' 'once a month,' 'once a quarter,' and 'once a year.' When a friend gets married, you get knocked down a level of frequency. When they have a first child, you're knocked down again. Same with the second child. If they have a third kid, you're off the charts completely."

Psychologist Terri Apter says that time was in fact the main luxury on display in the *Sex in the City* series. "It was appealing not just for the sex and the clothes, but for the image of a group of women who had ample time for each other. That was the nub of the fantasy."

MONEY, LIKE TIME, CAN BE a complicating factor. Those who dispense friend advice often discourage moneylending, yet Hurschka found that Americans lend money to friends quite regularly. Spending money on others can also be awkward and problematic: Even if it's greatly appreciated, it can be distressing on the receiving end, in that it's hard to "make up for it" if one doesn't have decent cash flow herself. Jacob, who earns a high salary, loves to treat others but sometimes goes out of his way to do it discreetly, to save them any discomfort. "Once at a restaurant, I pretended I lost a bet to my fiancée and had to buy her and her friends dinner as a result." His grandparents, for whom he has great admiration, always took people out to dinner, and Jacob likes carrying on the tradition. "Also, I feel like I lucked into my job, so while I don't want to be irresponsible with my money, it would feel wrong to hoard something I don't feel I necessarily earned, especially since other friends treated me back when I was penniless. And it gives me happiness, so there's a selfish aspect to it."

An ad executive in his fifties related that since the recession, he's seen friend problems bubble up over money. "When men are doing badly financially, they retreat. Friendships with them become more complicated. They don't want to admit they are suffering, and they can't easily talk from a perspective of hope the way a younger man would, even if he were also doing badly. I thought friendship would be easier at my age since I've known many of these guys for a long time, but I feel more 'stress fractures' in the relationships now than I did when we were all going out on each other's expense accounts and having a wonderful time." It's a doubly unfortunate pattern since these men are withdrawing from buddies right when they could really use the support that friendship provides.

As Cacioppo pointed out, feeling happiness for a friend's high points is the mark of a strong connection. That doesn't mean we won't also have flashes of envy, however. The classic advice is to pay attention to envy as a signal to improve a corresponding area of your own situation; it's been proved true time and time again. I also think a direct approach can clear the air. Years ago, a few days after graciously receiving the news that a friend was pregnant, I saw her and had a different reaction: I burst into tears and confessed a more complex emotion than vicarious delight. "I'm so happy for you," I said, "but I'm afraid I'll be envious of you and that it will affect our friendship." "I'm afraid I'll be envious of you and your single life!" she answered. To hear her understanding of the darker side of my feelings, as well as the admission that she herself was having mixed feelings, completely drained the potential negativity out of the moment and made us even closer.

Jan Yager shares a tactic in *When Friendship Hurts* for warding off a friend's envy of you: "Be careful about saying things to your friend who did not get a promotion or some other goal that she is aiming for but that you just achieved, in the guise of making your friend feel better. Instead of building up your friend's self-esteem the comments unwittingly focus on her unfulfilled dreams and goals. If possible, keep what you have achieved separate from what your friend is still striving for."

Carrying out favors for friends can also be a tricky business. "It's more awkward for me when a really good friend asks me for a work-related connection or favor than if a less close friend does," says Dalton Conley. "With the former there is intense pressure because of the friendship, but now it's shifted into the world of business, where the norm should be meritocracy. Even if someone is literally my best friend, it doesn't mean I want to stand behind their writing or their research and attach my name and recommendation to it."

Conversely, not asking friends for favors can be offensive. Hruschka writes of a study in Japan and the United States in which college students "were asked how they would feel if a close friend needed help—with taking care of a dog, fixing a computer, or having a place to stay for the night—but didn't ask them for help, going rather to another friend or a market-based service. In both cultures, students said they would feel sadder, more disappointed, and less close with the friend than they would have if the friend had come to them for help."

THERE'S SOMETHING BOTH COMFORTING and unsettling about reading ancient guidance on friendship. Its resonance with modern life is a clear clue that this relationship is central to our core selves. Yet it makes one wonder if we humans have made any progress at all, if all those intervening years of work in psychology and philosophy have not yielded much better advice. Consider these gems from the ancient Roman Cicero: "The pains friendship may cause are not a reason to avoid friendships. Friendships made for ulterior motives sacrifice the joy of friendships made out of pure affection. Friendship cannot be bought. The demands of friends must be watched carefully—sometimes friendships must be broken; sometimes this happens due to a gradual change in character. Friendship should aid virtue, not vice. The really virtuous are willing to hear the truth from their friends."

Cicero's ideals still hold, but you don't have to be an ideal friend to be a good one or to get the most out of friendfluence. People pleasers and martyrs have their own sets of problems and don't necessarily get more out of their friendships for sacrificing themselves. For most of us who could stand to be more generous and more forgiving, however, generating positive friendfluence starts with personal resolve: As you shield yourself from isolation and from pals who are poor influences, you must also be more cognizant of your special role as a friend to others. Being a friend is a great honor and responsibility, so treat your friends carefully. Friends who are mutually aware of their impact on each other can work to stay close and help each other get the most out of life.

Jacob and one of his college buddies had a "half-life" celebration dinner when they realized they'd been friends longer than they hadn't been friends. It was a nicely improvised occasion where no standard ritual exists, unfortunately. I'm not one for creating yet another consumer holiday, but isn't it telling that we don't usually mark the length of our close friendships the way we do romantic anniversaries? The "matrimania" in our culture that has bred over-the-top weddings, renewals of wedding vows, Valentine's Day splurges, and so forth, which the foremost singles' advocate and sociologist Bella DePaulo has documented and critiqued, could surely use counterbalancing with celebrations of other kinds of relationships that are central to us.

That brings us from a personal quest to be a good friend to cultural attitudes about friendship. What would a society that truly recognized friendfluence look like? For one thing, it wouldn't put everything on the individual. Shoring up your willpower is a noble goal, but it's very difficult. Such attempts don't usually work, or we wouldn't have yet another crop of books about procrastination, dieting, goal setting, and so on. Acknowledging our ability to help one another in these areas might make us more successful in meeting our goals. Nature does not want you to finish a paper for school or lose weight per se. She could care less. She built you to fit in with your "tribe," though. So, if you befriend those who are already accomplishing what you've been independently struggling to achieve, your habits will more easily converge with theirs. And aside from all of the subtle behavioral and moral influences we've unearthed, direct help from friends is also an underutilized resource. Ask more of your friends and you'll give them the gift of feeling influential.

In order to fully tap the power of friendship, we should collectively spend less time emphasizing the differences between male and female friendships and more time acknowledging that everyone could benefit from stronger friendships—not the idealized pop cultural versions, but the real, sometimes annoying, sometimes troubling, and sometimes transcendent kind.

Parents of little kids should be told that social skills and the ability to sustain friendships might be more important than the learning that occurs through all those structured activities that don't let kids plan and sort things out among themselves. Parents of teens should get the message that the peer group will often trump them and that it's better to invite the gang over to strengthen ties rather than criticize their sons' or daughters' choices. A friendless child should get extra help from parents and teachers in getting off the downward spiral and onto the upward spiral of social connection. Adults in romantic relationships should value their own time with friends without guilt. Couples should cultivate mutual friends, too, and turn to others for support, fun, and inspiration, rather than put everything on a spouse. Singles should not be automatically and judgmentally viewed as free-floating and disconnected, but as people who are very likely to have one or more "significant others" in their lives—of the friend variety.

Gay communities first taught interested researchers in the '70s and '80s about the great potential of friendship for those who lacked familial and neighborly support. Today the growing number of people who are living alone (whether gay, straight, divorced, widowed, in a romantic relationship, hoping to be in one, or content not to be in one) is likewise showcasing friends' often underestimated ability not just to accompany us through life but to make us who we are and to make us better than we'd ever be without them. Our culture seems primed to fully recognize how friendship both reflects and alters our natures. In the meantime, hold your own friends up high, where they belong.

EPILOGUE

MY FRIENDS HAVE FOR SO LONG meant so much to me—that hasn't changed since I began writing this book—but my appreciation for and understanding of them have certainly come into sharper focus this past year. For those I barely even keep in touch with: I realized I can still enjoy our friendship, here and now, through both the power of nostalgia and the knowledge that what I got out of learning your life stories and sharing experiences with you is still within me somewhere, in the way I act and look at the world. For those with whom I touch base only via social media: How nice it is to be a little virtual fly on the wall, catching glimpses of how your lives are progressing. For closer friends whom I see and interact with all the time: Even if I don't feel like going out or picking up the phone sometimes, it never, ever fails—talking to you raises my spirits, energizes me, makes me laugh, and helps me sort out any struggles while simultaneously allowing me to escape them for a while.

Whereas I might have separated social life and work somewhat in my mind before, I've since learned to see my own and other people's career trajectories in terms of the friends who surround them, not just in terms of their own choices or talents. It's like when you learn a new word and suddenly stumble upon it everywhere: Once you realize how much friends influence each other, you start to see their impact on yourself and others you know, and you're quicker to attribute people's circumstances to their friends as well as their backgrounds or parents or inner qualities.

I had the experience recently of realizing, after several lopsided efforts to connect, that a long-term friend whom I adore is simply not interested in being my friend anymore (or is perhaps withdrawing from all of her loved ones. As with many friend falling-outs, it is hard to precisely chart the demise). It hurts. But the credo of this book—to be more conscious of how friends are affecting you—has helped me accept the loss and turn my attention to those who seek me out.

Taken together, the many interesting facts about friendship I've learned have convinced me that there is real fulfillment to be gained from integrating friends fully into one's life: to view friends not as bonus features but as key relationships. In light of this conviction, I got extra joy at a recent occasion during which my family and friends had an opportunity to mingle more than usual. If my family members really know my friends, and vice versa, they can all really know me. And isn't that something we're all after: to be truly known by others?

While reporting and researching, I frequently discussed insights I'd learned from experts or storytellers with my husband, who moved here to the United States from Mexico in 2007. Of all the reflections I've casually collected from people about their friendships (everyone has one), it was his that hit me hardest: The most difficult part of immigrating for him—harder than mastering the language, or starting a career nearly from scratch in his thirties, or moving away from the immediate and extended family he had lived with his entire life—was leaving his close, long-term friends. Because he's gregarious and has made buddies here, and because he simply hadn't expressed his feelings on the matter before, it had never dawned on me just how trying that aspect of the transition was and how he had sacrificed his own friendships to be with me, ironically someone intent on preaching the virtues of friendship.

I look forward to watching my husband get closer to his new friends and to strengthening our mutual friendships. And because I'm a few weeks away from giving birth to a boy, I'd like to send out a maternal wish forged in social science: I don't care too much if my son is smart or strong or handsome; I just want him to be friendly.

ACKNOWLEDGMENTS

I WANT TO HEARTILY THANK the academics, clinicians, and journalists whose research, theories, and experiences make up much of the raw material of this book. I'm equally grateful to all who shared their personal friendship stories with me. Thank you for your time and candor.

I am incredibly grateful to two brilliant women who were essential to this project: Melissa Danaczko and Gillian MacKenzie.

Melissa: Thanks so much for all of your excellent input and revisions and for being so very kind, attentive, and helpful every step of the way. You are the ideal editor. Your friends are extremely lucky to have you, and I hope I join their ranks.

Gillian: Thanks for being such a wonderful and effective advocate throughout, and for your support for this idea from its earliest stages. As my friend as well as my agent, your energy and passion for ideas have had a great influence on me.

For their expertise and hard work, lots of thanks to Joe Gallagher, Nora Reichard, Kathryn Santora, and the rest of the team at Doubleday.

I'm very grateful to the past and present editors at *Psychology Today* for lessons in social-science journalism. I am especially indebted to Kaja Perina, for being such a generous mentor, and Lybi Ma, that friend at work who boosts well-being and productivity. Jo Colman: Thanks for motivating me by modeling admirable levels of self-discipline.

Thanks to Ellen and Jackie (and the other Armstrongs and

Groots) for encouragement. And *gracias* to the Escaleras for always asking, *"Como va el libro?"*

Denise and Joseph Flora: Thank you for enabling and tolerating all kinds of childhood creative endeavors, including writing. Your enthusiasm for *Friendfluence* is very much appreciated! A special thanks to Denise for her loving help with Caetano as I edited the book.

Thanks very much to cousins who have the elevated title of "friend": Sara Ridder, Sasha Rene, and Danielle Vorbe; and to three very supportive aunts: Ann Falkenberry, Nancy Fersh, and Kathy Treat.

To my friends: Thank you from the bottom of my heart. Some of you shared your friend tales, some let me tell tales of you. Those who do not directly appear in the book definitely influenced the observations therein. In case it isn't obvious, the happiness you have given me is what inspired me to take up the subject in the first place.

Special thanks to Nima Afshar, Carmen and Dan Brenner, Nardiz Cooke, Madhavi Dandu, Lola Diaz, Christina and Leo Drogaris, Marcos Fajardo, Jessica and Rick Greenberg, Carlos Hernandez, Evan Hughes, Maggie Janes-Lucas, Jesse Levine, Mariana Lopez, Carlos Lopez de Alba, Joe McCarthy, Kat McGowan, Michelle Orange, Molly Pulda, Jana Prikryl, Amy Rosenberg, Bianca Sandovai, Susan Sokirko, Stacey Vanek Smith, Meline Toumani, and Seth Wenig. Extra special thanks to Erika Schimik and Erika Strand.

I'd like to thank two friends I long ago fell out of touch with who nonetheless continue to influence me in a very positive way: Amy Corning and Ariadne Daskalakis.

To my kindred spirit Sofia Jozefowicz: How glad I am that you knocked on door 301 in 1993! Chris Jozefowicz: You're the best third wheel two BFFs could want. Thank you both for two decades of fun and devotion.

Much gratitude to my dear friend and neighbor Dawn Siff for extraordinary support before and after the major life change that coincided with this project.

For editorial suggestions on the book's proposal and years of affectionately dispensed career (and life) advice and brotherly friendship, thank you very much, Gary Sernovitz.

A big thanks to Matt Hutson for editorial suggestions and for setting such a great book-writing example.

From the first draft of the proposal to the moment after I turned in the last proof of the manuscript, I have been a constant beneficiary of one friend's editorial acumen, deep understanding of people and relationships, and caring reassurances. Adelle Waldman: I am so fortunate to have you as a friend, and I truly can't thank you enough.

My sweet son, Caetano: Thank you so much for being a comforting silent companion while I wrote much of this book and an inspirational (and adorably noisy!) companion as I revised it.

Finally, I am supremely grateful for Giovanni Escalera's faith in me and enveloping love: You are the light that never goes out.

NOTES

INTRODUCTION

3 **The median age of first marriage:** The U.S. Census Bureau's 2010 American Community Survey.

3 **One hundred million Americans:** Tara Parker-Pope, "In a Married World, Singles Struggle for Attention," *New York Times*, September 19, 2011.

4 **Klinenberg blows apart the stereotype:** Eric Klinenberg, "One's a Crowd," *New York Times*, February 4, 2012.

7 **One study of nurses:** Tara Parker-Pope, "What Are Friends For? A Longer Life," *New York Times*, April 20, 2009.

CHAPTER 1

9 **"It had gotten scarier and scarier":** E-mail and phone interviews with Holly Kile, February 23, 2011, and March 2, 2011.

10 **Friendships are the least institutionalized:** R. Adams, R. Blieszner, and B. de Vries, "Definitions of Friendship in the Third Age: Age, Gender, and Study Location Effects," *Journal of Aging Studies* 14 (2000): 117–33.

11 ***"How can life be worth living"*****:** Cicero, "De Amicitia," trans. Andrew P. Peabody (Boston, MA: Little, Brown, and Company, 1884).

11 ***"You paint my life brighter"*****:** Hallmark.com.

11 **The French essayist Michel Eyquem de Montaigne:** Michel de Montaigne, *Montaigne: Essays,* trans. John M. Cohen (New York: Penguin Books, 1993).

12 **Biological anthropologist Gwen Dewar, Ph.D.:** Gwen Dewar, "Should Parents Be Friends with Their Kids?," Parentingscience.com, September 9, 2009.

12 **Some researchers believe that parents:** Jean M. Twenge and W. Keith Campbell, *The Narcissism Epidemic: Living in the Age of Entitlement* (New York: Free Press, 2009).

12 **Perhaps the sharpest distinction Montaigne:** Montaigne, *Montaigne: Essays*.

14 **By doing so, he has mapped out:** Phone interview with Brian de Vries, March 2, 2011.

14 **A key study analyzed older adults:** Adams, Blieszner, and de Vries, "Definitions of Friendship in the Third Age," *Journal of Aging Studies* 14 (2000): 117–33.

15 **Less frequently mentioned:** Ibid.

17 **"Depending on the culture":** Daniel J. Hruschka, *Friendship: Development, Ecology, and Evolution of a Relationship* (Berkeley: University of California Press, 2010).

18 **The females recently studied:** Gerald Kerth, Nicolas Perony, and Frank Schweitzer, "Bats Are Able to Maintain Long-Term Social Relationships Despite the High Fission–Fusion Dynamics of Their Groups," *Proceedings of the Royal Society*, February 9, 2011.

18 **Gerald Kerth, Ph.D., the head:** Jennifer Viegas, "Animals Make Friends, Too," *Discovery News*, February 8, 2011.

18 **A recent study of macaque primates:** Jerome Micheletta and Bridget Waller, "Friendship Affects Gaze Following in a Tolerant Species of Macaque, Macaca Nigra," *Animal Behaviour* 83, no. 2 (2012): 459–67.

19 **"Humans putting their heads together":** Michael Tomasello, *Why We Cooperate* (Cambridge, Mass.: MIT Press, 2009).

19 **Daniel Hruschka writes that:** Hruschka, *Friendship*.

20 **This may be an explanation for what:** Ibid.

20 **Steve Weitzenkorn, now sixty:** E-mail and phone interviews with Steve Weitzenkorn, March 3, 2011.

21 **But Geoffrey Greif, D.S.W.:** Geoffrey Greif, *Buddy System: Understanding Male Friendships* (New York: Oxford University Press, 2008).

22 **Greif interviewed about five hundred people:** Geoffrey Greif, "Men and Women Have Different Styles of Same Sex Friendships," Psychologytoday.com, October 27, 2009.

22 **In spite of these surface differences:** Hruschka, *Friendship*.

23 **received a new label:** Jennifer 8. Lee, "The Man Date," *New York Times*, April 10, 2005.

24 **In 2009 the University of Chicago instituted:** Dave Newbart, "Men, Women Share U. of C. Rooms," *Chicago Sun-Times*, July 6, 2009.

24 **Lucy Taylor, a British journalist:** Lucy Taylor, "Why Women Need Male Friends," *Daily Mail* (UK), September 1, 2009.

25 **a fellow British journalist, Sarfraz Manzoor:** Sarfraz Manzoor, "Why I Hang Out with Girls. It's Not What You Think," *Observer* (UK), June 6, 2009.

26 **Young adults seem especially comfortable:** Pamela Paul, "We're Just Friends, Really!," *Time,* September 1, 2003.

26 **a study that explored the role of friends:** K. S. Burditt and T. C. Antonucci, "Relationship Quality Profiles and Well-Being

Among Married Adults," *Journal of Family Psychology* 21, no. 4 (December 2007): 595–604.

26 **conducted their own qualitative survey:** Suzanne Degges-White and Christine Borzumato-Gainey, *Friends Forever: How Girls and Women Forge Lasting Relationships* (Plymouth, UK: Rowman & Littlefield, 2011).

27 **One study in fact found that both men and women:** Camille Chatterjee, "Can Men and Women Be Friends?," *Psychology Today* (September 1, 2001).

27 **In 2000, Penn State researchers:** Walid A. Afifi and Sandra L. Faulkner, "On Being 'Just Friends': The Frequency and Impact of Sexual Activity in Cross Sex Friendships," *Journal of Social and Personal Relationships* 17, no. 2 (April 2000): 205–22.

27 **In 2010, Justin Lehmiller, Ph.D., and colleagues:** Justin L. Lehmiller, Laura E. Vanderdrift, and Janice R. Kelly, "Sex Differences in Approaching Friends with Benefits Relationships," *Journal of Sex Research* 47 (2010): 1–10.

28 **the mere act of having sex can trigger romantic feelings:** Helen Fisher, *Why We Love: The Nature and Chemistry of Romantic Love* (New York: Henry Holt, 2004).

28 **When Peter Nardi, Ph.D., professor emeritus:** Phone interview with Peter Nardi, February 22, 2011.

29 **More recent work from de Vries:** Brian de Vries and David Megathlin, "The Dimensions and Processes of Older GLBT Friendships and Family Relationships," *Journal of GLBT Family Studies* 5 (2009): 82–98.

29 **Shane Allen, twenty-four, and Felipe Baeza, twenty-three:** Interview with Shane Allen and Felipe Baeza, March 6, 2011.

31 **Yet, as he writes in one of his papers:** Brian de Vries, "Friendship and Family: The Company We Keep," *Transition Magazine* 40, no. 4 (Winter 2010).

31 **Consider the health care system:** Phone interview with Brian de Vries, March 2, 2011.

32 **"As a culture, we're less likely":** de Vries, "Friendship and Family."

CHAPTER 2

35 **Half a century ago:** Suzanne Degges-White and Christine Borzumato-Gainey, *Friends Forever: How Girls and Women Forge Lasting Relationships* (Plymouth, UK: Rowman & Littlefield, 2011).

35 **The mere-exposure effect lends nearness:** R. B. Zajonc, "Mere Exposure: A Gateway to the Subliminal," *Current Directions in Psychological Science* 10, no. 6 (December 2001): 224–28.

35 **You'll give off a better first impression:** Simon M. Laham, Peter Koval, and Adam L. Alter, "The Name-Pronunciation Effect: Why People Like Mr. Smith More than Mr. Colquhoun," *Journal of Experimental Social Psychology* 48, no. 3 (May 2012): 752–56.

35 **suburban sprawl, which puts people farther away:** Robert D. Putnam, *Bowling Alone: The Collapse and Revival of American Community* (New York: Simon & Schuster, 2000).

36 **Major life events and changes:** Degges-White and Borzumato-Gainey, *Friends Forever*.

36 **Back in 1937:** Daniel J. Hruschka, *Friendship: Development, Ecology, and Evolution of a Social Relationship* (Berkeley: University of California Press, 2010).

37 **Once we do disclose a feeling:** Degges-White and Borzumato-Gainey, *Friends Forever*.

37 **As with so much in psychology:** Ibid.

38 **As Hruschka puts it:** Hruschka, *Friendship*.

38 **One way to deepen:** Ibid.

38 **The longer you are friends with someone:** Ibid.

39 **"It's very nice to have someone who really gets you":** Frank Bruni, "The Sidekick No More," *New York Times*, March 18, 2011.

39 **"The level of similarity":** Degges-White and Borzumato-Gainey, *Friends Forever*.

40 **Jennifer Bosson, a psychologist:** Paul Kix, "Hating the Same Things," *New York*, March 27, 2011.

40 **Using data showing subjects' friendship ties:** James Fowler et al., "Correlated Genotypes in Friendship Networks," *Proceedings of the National Academy of Sciences* 108, no. 5 (2011): 1993–97.

40 **Hens' feathers change depending:** Michael Alvarez, "Genes and Social Networks," Psychologytoday.com, February 14, 2011.

41 **They tracked the number of e-mails:** David Marmaros and Bruce Sacerdote, "How Do Friendships Form?," NBER Working Paper No. 11530. Issued in August 2005.

42 **As one possible explanation:** Ibid.

42 **Three common explanations:** Degges-White and Borzumato-Gainey, *Friends Forever*.

43 **In the late '90s:** J. Tooby and L. Cosmides, *Friendship and the Banker's Paradox: Other Pathways to the Evolution of Adaptations for Altruism* (Oxford, UK: Oxford University Press, 1996).

43 **In fact, a few experiments:** Hruschka, *Friendship*.

44 **Consider the friend niche limitation model:** Tooby and Cosmides, *Friendship and the Banker's Paradox: Other Pathways to the Evolution of Adaptation for Altruism,* in *Evolution of Social Behaviour Patterns in*

Primates and Man, ed. W. G. Runciman, J. Maynard Smith, and R. I. M. Dunbar. *Proceedings of the British Academy* 88 (1996): 119–43.

44 **Tooby and Cosmides suggest that we evolved:** Ibid.

44 **You've also evolved to want:** Ibid.

45 **"If human friendship is strategic":** P. DeScioli and R. Kurzban, "The Company You Keep: Friendship Decisions from a Functional Perspective," in *Social Judgment and Decision Making,* ed. J. I. Krueger (New York: Psychology Press, 2011).

46 **"You want allies before the dispute":** Phone interview with Peter DeScioli, March 25, 2011.

46 **He collected a sample of 11 million:** Peter DeScioli, Robert Kurzban, Elizabeth N. Koch, and David Liben-Nowell, "Best Friends: Alliances, Friend Ranking, and the MySpace Social Network," *Perspectives on Psychological Science* 6, no. 1 (2011): 6–8.

47 **"My friend got engaged in July":** From TheKnot.com message boards.

47 **"Have any of you experienced":** Ibid.

48 **a typical reaction of those:** Phone interview with Peter DeScioli, March 25, 2011.

49 **Psychologist Carolyn Weisz, Ph.D.:** Phone interview with Carolyn Weisz, April 1, 2011.

50 **Or take the real-life example:** Interview with Solomon Dumas and Slim Mello, April 26, 2011.

52 **Weisz asked a group of college freshmen:** Carolyn Weisz and Lisa F. Wood, "Social Identity Support and Friendship Outcomes: A Longitudinal Study Predicting Who Will Be Friends and Best Friends 4 Years Later," *Journal of Social and Personal Relationships* 22, no. 3 (June 2005): 416–32.

52 **"When one friend gets into exercise":** Phone interview with Carolyn Weisz, April 1, 2011.

52 **"In effect we have five intimate friends":** E-mail interview with Robin Dunbar, April 11, 2011.

53 **Some individuals might be particularly:** J. Powell, P. A. Lewis, N. Roberts, M. Garcia-Finana, and R. I. M. Dunbar, "Orbital Prefrontal Cortex Volume Predicts Social Network Size: An Imaging Study of Individual Differences in Humans," *Proceedings of the Royal Society B: Biological Sciences,* 2012.

53 **But another study of Dunbar's:** Sam G. B. Roberts, Robin I. M. Dunbar, Thomas V. Pollet, and Toon Kuppens, "Exploring Variation in Active Network Size: Constraints and Ego Characteristics," *Social Networks* 31, no. 2 (May 2009): 138–46.

53 **Another team more recently found:** Sian Beilock, "When It Comes to Our Social Networks, Brain Size Matters," Psychologytoday.com, February 25, 2011.

54 **"If anything":** E-mail interview with Robin Dunbar, April 11, 2011.

55 **A study by a Dutch sociologist:** G. W. Mollenhorst, "Networks in Contexts. How Meeting Opportunities Affect Personal Relationships," Utrecht University, 2009. ICS Dissertation Series, vol. 150.

56 **The year was 1995:** Google's company history at google.com.

56 **The FDR Presidential Library and Museum:** The "Special Relationship": Churchill, Roosevelt and the Emergence of the Anglo-American Alliance, 1939–1945, British Diplomatic Files. FDR Presidential Library & Digital Archives, http://docs.fdrlibrary.marist.edu/anglo.html.

57 **Albert Speer, an architect:** Wesley Yang, "Hitler's Best Friend," Salon.com, September 26, 2002.

CHAPTER 3

58 **"She was so happy":** Phone interview with Suzanne Ludlum, July 5, 2011.

59 **As the prominent child development psychologist:** Kenneth H. Rubin and Andrea Thompson, *The Friendship Factor: Helping Our Children Navigate Their Social World—and Why It Matters for Their Success and Happiness* (New York: Penguin, 2002).

60 **Peter Gray, Ph.D., a professor of psychology:** Malcolm Gladwell, "Annals of Behavior: Do Parents Matter?," *New Yorker,* August 17, 1998.

60 **"In children's literature, the character often has adventures":** Phone interview with Philip Nel, July 22, 2011.

60 **Friendships sprout much earlier:** William M. Bukowski, Andrew F. Newcomb, and Willard W. Hartup, eds., *The Company They Keep: Friendships in Childhood and Adolescence* (Cambridge, UK): Cambridge University Press, 1996.

61 **In one study, the first time:** Ibid.

61 **Toddlers who entered a new:** Ibid.

61 **Little kids are on the lookout:** Rubin and Thompson, *The Friendship Factor.*

61 **Once they reach the age:** Ibid.

61 **They talk and seem:** Bukowski, Newcomb, and Hartup, *The Company They Keep.*

62 **Kids may exchange fairly:** Ibid.

62 **A study from the '80s:** Ibid.

62 **Just like grown-ups:** Ibid.

62 **Rubin did find that if children:** Phone interview with Kenneth Rubin, July 6, 2011.

62 **How exactly do little pals:** Bukowski, Newcomb, and Hartup, *The Company They Keep.*

62 **Drew, a sunny eight-year-old boy:** Phone interview with Drew, July 7, 2011.

63 **Of course, mutual affection must be present:** Rubin and Thompson, *The Friendship Factor.*

63 **While there are exceptions:** Ibid.

63 **When I asked Ella:** Skype interview with Ella, June 22, 2011. ("Jessica" is a pseudonym for Ella's friend.)

63 **in the United States and Europe:** Daniel J. Hruschka, *Friendship: Development, Ecology, and Evolution of a Social Relationship* (Berkeley: University of California Press, 2010).

64 **"Before a child is walking and talking":** Rubin and Thompson, *The Friendship Factor.*

64 **It's unfair, but looks matter:** Ibid.

65 **Unsurprisingly, they're able to make:** Ibid.

65 **Children who are natural social stars:** Ibid.

65 **They tend to be unable to appreciate:** Ibid.

65 **An anxious or withdrawn child:** Phone interview with Kenneth Rubin, July 6, 2011.

66 **Some children who are socially isolated:** William M. Bukowski, Brett Laursen, and Betsy Hoza, "The Snowball Effect: Friendship Moderates Escalations in Depressed Affect Among Avoidant and Excluded Children," *Development and Psychopathology* (October 2010): 749–57.

66 **Finding a best friend who:** Rubin and Thompson, *The Friendship Factor.*

67 **A Canadian study of kids:** Andre Picard, "Don't Shut Disabled Kids Out of Society," *Globe and Mail*, January 30, 2012.

67 **A caveat is in order:** Bukowski, Newcomb, and Hartup, *The Company They Keep.*

68 **As for kids:** Kenneth Rubin, Bridget Fredstrom, and Julie Bowker, "Future Directions in . . . Friendship in Childhood and Early Adolescence," *Social Development* (February 14, 2008).

68 **You may recall that around ages:** Ibid.

69 **When she is disagreeing with a pal:** Bukowski, Newcomb, and Hartup, *The Company They Keep.*

69 **I spoke to five gregarious girls:** Interview with girls in Columbia, South Carolina, May 3, 2011.

69 **As friendships are embedded:** Michael Thompson, Lawrence J. Cohen, and Catherine O'Neill Grace, *Best Friends, Worst Enemies: Understanding the Social Lives of Children* (New York: Ballantine Books, 2001).

70 **Astrid, now thirty-five:** E-mail interview with Astrid, whose name has been changed, July 11, 2011.

71 **Best friends do seem to have:** Hilary Stout, "A Best Friend? You Must Be Kidding," *New York Times*, June 16, 2010.

71 **Ella, the ten-year-old:** Skype interview with Ella, June 22, 2011.
72 **Friendless children are lonelier than other:** Bukowski, Newcomb, and Hartup, *The Company They Keep.*
72 **They are also likely to grow:** Ibid.
72 **Though the thought of an innocent:** Ibid.
72 **But even kids who:** Ibid.
73 **Preschoolers who are securely:** Ibid.
73 **On the flip side:** Ibid.
73 **Summoning Goldilocks, researchers:** Ibid.
73 **Divorce is linked to social:** Ibid.
73 **Friends trump siblings:** Brenda L. Volling, Lise Youngblade, and Jay Belsky, "Young Children's Social Relationships with Siblings and Friends," *American Journal of Orthopsychiatry* 67, no. 1 (January 1997): 102–11.
74 **Kids who are good friends:** Laurie Kramer and Amanda Kowal, "Sibling Relationship Quality from Birth to Adolescence: The Enduring Contributions of Friends," *Journal of Family Psychology* 19, no. 4 (December 2005): 503–11.
74 **John Coie, Ph.D., of Duke University:** Thompson, Cohen, and O'Neill Grace, *Best Friends, Worst Enemies.*
75 **Four percent of kids:** Ibid.
75 **"It is done through thousands of little":** Ibid.
76 **some kids might literally be born:** Nicholas A. Christakis and James H. Fowler, *Connected: The Surprising Power of Our Social Networks and How They Shape Our Lives* (New York: Little, Brown and Company, 2009).
76 **"Everyone at our school":** Interview with girls in Columbia, South Carolina, May 3, 2011.
76 **"Something like learning karate":** Phone interview with James Olsen, July 5, 2011.
77 **Friendship is different from popularity:** Bukowski, Newcomb, and Hartup, *The Company They Keep.*
77 **Having a friend melts away loneliness:** Ibid.
77 **Rubin makes a distinction:** Phone interview with Kenneth Rubin, July 6, 2011.
77 **journalist Alexandra Robbins makes the case:** Alexandra Robbins, *The Geeks Shall Inherit the Earth* (New York: Hyperion, 2011).
77 **Grace and her friends:** Interview with girls in Chapin, South Carolina, May 3, 2011.
78 **As Thompson put it:** Thompson, Cohen, and O'Neill Grace, *Best Friends, Worst Enemies.*
78 **When Christine, now thirty-four:** Interview with Christine, June 4, 2011. (Her name and other names are changed.)

80 **"friendships aren't always symmetrical":** Phone interview with Kenneth Rubin, July 6, 2011.

80 **When he asked several hundred:** James P. Olsen, Gilbert R. Parra, Robert Cohen, Corrie L. Schoffstall, and Clayton Joe Egli, "Beyond Relationship Reciprocity: A Consideration of Varied Forms of Children's Relationships," *Personal Relationships* 19, no. 1 (March 2012): 72–88.

80 **"Children who are believed":** Phone interview with James P. Olsen, July 5, 2011.

81 **Other enemies become:** Phone interview with Maurissa Abecassis, July 11, 2011.

82 **Once you remove those:** Benedict Carey, "Can an Enemy Be a Child's Friend?," *New York Times*, May 17, 2010.

83 **One recent survey of third graders:** Pamela Paul, "The Playground Gets Even Tougher," *New York Times*, October 8, 2010.

83 **Such behavior starts early:** Phone interview with Maurissa Abecassis, July 11, 2011.

83 **"Our moms tell us not to gossip":** Interview with girls in Chapin, South Carolina, May 3, 2011.

84 **A recent study shows that while it's true:** N. A. Card, B. D. Stucky, G. M. Sawalani, and T. D. Little, "Direct and Indirect Aggression During Childhood and Adolescence: A Meta-Analytic Review of Gender Differences, Intercorrelations, and Relations to Maladjustment," *Child Development* 79, no. 5 (2008): 1185–1229.

85 **Ella's story bears witness:** E. V. E. Hodges, M. J. Malone, and D. G. Perry, "Individual Risk and Social Risk as Interacting Determinants of Victimization in the Peer Group," *Developmental Psychology* 33 (1997): 1032–39.

85 **Friendless victims do show:** E. V. E. Hodges and D. G. Perry, "Personal and Interpersonal Antecedents and Consequences of Victimization by Peers," *Journal of Personality and Social Psychology* 76 (1999): 677–85.

85 **A Canadian study of cortisol:** R. E. Adams, J. B. Santo, and W. M. Bukowski, "The Presence of a Best Friend Buffers the Effects of Negative Experience," *Developmental Psychology* 47 (2011): 1786–91.

CHAPTER 4

87 **It was the '80s and Lydia:** Interview with Lydia, August 9, 2011. (Names of Lydia and Rachel have been changed.)

91 **In fact, to the average:** Kenneth H. Rubin and Andrea Thompson, *The Friendship Factor: Helping Our Children Navigate Their Social World—and Why It Matters for Their Success and Happiness* (New York: Penguin, 2002).

91 **While 9 percent recounted:** Judith Rich Harris, *The Nurture Assumption: Why Children Turn Out the Way They Do* (New York: Free Press, 2009).

91 **As children enter middle school:** Michael Thompson, Lawrence J. Cohen, and Catherine O'Neill Grace, *Best Friends, Worst Enemies: Understanding the Social Lives of Children* (New York: Ballantine Books, 2001).

92 **Still, it's all a continuum:** Ibid.

92 **As Kenneth Rubin describes it:** Rubin and Thompson, *The Friendship Factor.*

92 **For adolescents who have antagonistic:** Kenneth H. Rubin, Kathleen M. Dwyer, Cathryn Booth-LaForce, Angel H. Kim, Kim B. Burgess, and Linda Rose-Krasnor, "Attachment, Friendship, and Psychosocial Functioning in Early Adolescence," *Journal of Early Adolescence* 24, no. 4 (November 2004): 326–56.

93 **Great friends can also teach:** Rubin and Thompson, *The Friendship Factor.*

93 **James Olsen says that while:** Phone interview with James P. Olsen, July 5, 2011.

94 **By adolescence, teens have settled:** Phone interview with Carl Pickhardt, August 22, 2011.

94 **Just having those few:** William M. Bukowski, Andrew F. Newcomb, and Willard W. Hartup, eds., *The Company They Keep: Friendships in Childhood and Adolescence* (Cambridge, UK): Cambridge University Press, 1996.

94 **And if an adolescent starts:** Ibid.

94 **While Pickhardt is hopeful:** Shari Miller-Johnson, John D. Coie, Anne Maumary-Gremaud, John Lochman, and Robert Terry, "Relationship Between Childhood Peer Rejection and Aggression and Adolescent Delinquency Severity and Type Among African American Youth," *Journal of Emotional and Behavioral Disorders* 7, no. 3 (Fall 1999): 137–46.

96 **American adolescents are especially:** J. A. Wall, T. G. Power, and C. Arbona, "Susceptibility to Antisocial Peer Pressure and Its Relation to Acculturation in Mexican-American Adolescents," *Journal of Adolescent Research* 8, no. 4 (1993): 403–18.

96 **But as Daniel Hruschka points out:** Daniel J. Hruschka, *Friendship: Development, Ecology, and Evolution of a Social Relationship* (Berkeley: University of California Press, 2010).

96 **Half of American teenagers are sending:** Amanda Lenhart, "Teens, Cell Phones and Texting: Text Messaging Becomes Centerpiece Communication," Pew Internet and American Life Project, May 19, 2012, http://pewresearch.org/reports/2012/Teens-and-smartphones.aspx.

96 **Thirty-five percent of teens:** Ibid.

97 **teenage friendships have the distinction:** Shawn Amos, "High School Is the Place to Start a Band," May 18, 2009, GetBack Yahoo blog.

98 **A Dutch study confirms:** "Friendship Is Mainly About 'Me, Me, and Me,'" *Science Daily*, October 22, 2009.

98 **These projections can be very accurate:** A. J. Bahns and C. S. Crandall, *Prejudice Comes in Pairs: Friends Assort on Prejudices Without Discussion.* Manuscript in preparation, 2010.

98 **Jesse Rude, Ph.D., and Daniel Herda:** Jesse Rude and Daniel Herda, "Best Friends Forever? Race and the Stability of Adolescent Friendships," *Social Forces* 89, no. 2 (2010): 585–607.

98 **Rude and Herda were most interested:** Ibid.

99 **Meanwhile, Pickhardt sees:** Phone interview with Carl Pickhardt, August 22, 2011.

99 **"Here in the States if you go":** Skype interview with Alexys, August 20, 2011. (Chloe is a pseudonym for her friend.)

101 **Having friends who care about grades:** Bukowski, Newcomb, and Hartup, *The Company They Keep*.

101 **Surprisingly, though, in a study:** Marie-Helene Veronneau and Thomas J. Dishion, "Middle School Friendships and Academic Achievement in Early Adolescence: A Longitudinal Analysis," *Journal of Early Adolescence* 31, no. 1 (February 2011): 99–124.

101 **In another study:** Sarah E. Nelson and Thomas J. Dishion, "From Boys to Men: Predicting Adult Adaptation from Middle Childhood Sociometric Status," *Development and Psychopathology* 16 (2003): 441–459.

102 **Parents choose where their children live:** Judith Rich Harris, *The Nurture Assumption: Why Children Turn Out the Way They Do* (New York: Free Press, 2009).

102 **"By living in one neighborhood":** Ibid.

102 **In a study Harris recounts:** Ibid.

103 **Andy, seventeen, is a senior:** Interview with Andy, whose name has been changed, August 29, 2011.

103 **In fact, having deviant friends:** He Len Chung and Laurence Steinberg, "Relations Between Neighborhood Factors, Parenting Behaviors, Peer Deviance, and Delinquency Among Serious Juvenile Offenders," *Developmental Psychology* 42, no. 2 (2006): 319–31.

104 **One research team:** Ibid.

104 **When he was a doctoral student:** Greg Dimitriadis, *Friendship, Cliques, and Gangs: Young Black Men Coming of Age in Urban America* (New York: Teachers College Press, 2003).

104 **"There are two stereotypes":** Phone interview with Greg Dimitriadis, September 13, 2011.

105 **"self-directed kids are the ones":** Ibid.

106 **"The connection evolved from the mutual demands":** Ibid.

106 **"Jack Smith was my Bible studies":** Phone interview with Matt Hutson, August 15, 2011.

107 **"Matt is someone who is more":** Phone interview with John "Jack" Smith, August 19, 2011.

108 **We know that peer pressure:** Phone interview with Laurence Steinberg, August 29, 2011.

109 **Take the provocative term:** Timothy F. Piehler and Thomas J. Dishion, "Interpersonal Dynamics Within Adolescent Friendships: Dyadic Mutuality, Deviant Talk, and Patterns of Antisocial Behavior," *Child Development* 78, no. 5 (September/October 2007): 1611–24.

110 **If you catch a fourteen-year-old:** Phone interview with Laurence Steinberg, August 29, 2011.

110 **The more practice a kid has:** Piehler and Dishion, "Interpersonal Dynamics Within Adolescent Friendships."

110 **"Parents should know that their kid":** Phone interview with Laurence Steinberg, August 29, 2011.

110 **Steinberg and his team have recently:** Jason Chein, Dustin Albert, Lia O'Brien, Kaitlyn Uckert, and Laurence Steinberg, "Peers Increase Adolescent Risk Taking by Enhancing Activity in the Brain's Reward Circuitry," *Developmental Science* (March 2011): F1–F10.

111 **Correspondingly, they took fewer risks:** Ibid.

111 **A compounding effect:** Interview with Laurence Steinberg, August 29, 2011.

111 **If you're a woman, I'm sure you remember:** Sarah Kershaw, "Girl Talk Has Its Limits," *New York Times*, September 10, 2008.

111 **Amanda Rose, Ph.D., an assistant professor:** Phone interview with Amanda Rose, August 26, 2011.

112 **Having a friend with depression:** Mara Brendgen et al., "Links Between Friendship Relations and Early Adolescents' Trajectories of Depressed Mood," *Developmental Psychology* 46, no. 2 (March 2010): 491–501.

112 **A friend's engagement in:** Thomas J. Dishion and Jessica M. Tipsord, "Peer Contagion in Child and Adolescent Social and Emotional Development," *Annual Review of Psychology* 62 (January 2011): 189–214.

113 **You won't be shocked:** Ibid.

113 **But a recent social networking study:** Derek A. Kreager and Dana L. Haynie, "Dangerous Liaisons? Dating and Drinking Diffusion in Adolescent Peer Networks," *American Sociological Review* 76, no. 5 (October 2011): 737–63.

113 **The seemingly benevolent-sounding:** Dishion and Tipsord, "Peer Contagion in Child and Adolescent Social and Emotional Development."

114 **Grown-ups who keep an eye on kids:** Ibid.

114 **"The problem in adolescence is not":** J. P. Allen and J. Antonishak, "Adolescent Peer Influences: Beyond the Dark Side," in *Understanding Peer Influence in Children and Adolescents*, ed. M. J. Prinstein and K. A. Dodge (New York: Guilford Press, 2008), 141–60.

115 **Kids who participated in:** Ibid.

115 **Diana Marino, a thirteen-year-old who lives:** Phone interview with Diana Marino, August 19, 2011. (Elena is a pseudonym for her friend.)

117 **"A good friendship in adolescence is golden":** Phone interview with Carl Pickhardt, August 22, 2011.

CHAPTER 5

119 **For the past forty years:** Phone interview with Richard J. Levinson, January 4, 2012.

122 **Researchers at the University of Michigan:** O. Ybarra, P. Winkielman, I. Yeh, E. Burnstein, and L. Kavanagh, "Friends (and Sometimes Enemies) with Cognitive Benefits: What Types of Social Interactions Boost Executive Functioning?," *Social Psychological and Personality Science* 2, no. 3 (May 2011): 253–61.

122 **Creative ability is much harder:** Ruth Richards, *Everyday Creativity and New Views of Human Nature: Psychological, Social and Spiritual Perspectives* (Washington, D.C.: American Psychological Association, 2007).

122 **"Creativity is linking two ideas":** Phone interview with Michael Farrell, January 4, 2012.

124 **"When one of us dies":** Patricia Boccadoro, "Review of Matisse-Picasso Art and Archaeology Exhibitions," *Culturekiosque*, February 3, 2003.

124 **"For artists in general, and Picasso":** Phone interview with Elizabeth Cowling, February 25, 2011.

126 **"It's been the most rewarding":** Phone interview with James Fowler, December 8, 2011.

126 **Meliksah Demir, Ph.D., a professor:** Phone interview with Meliksah Demir, October 12, 2010.

127 **They found, controversially, that time with friends:** D. Kahneman, A. B. Krueger, D. A. Schkade, N. Schwarz, and A. A. Stone, "A Survey Method for Characterizing Daily Life Experience: The Day Reconstruction Method," *Science* 306 (December 3, 2004): 1776–80.

127 **Kahneman also found that commutes:** Ibid.

127 **Strong relationships, including those:** H. T. Reis and S. L. Gable, "Toward a Positive Psychology of Relationships," in *Flourishing: Positive*

Psychology and the Life Well-Lived, ed. Corey L. M. Keyes and Jonathan Haidt (Washington, D.C.: American Psychological Association, 2003), 129–59.

128 **In their network analyses:** From Nicholas A. Christakis and James H. Fowler, *Connected: The Surprising Power of Our Social Networks and How They Shape Our Lives* (New York: Little, Brown and Company, 2009).

128 **"Friendships built in religious congregations":** Chaeyoon Lim and Robert D. Putnam, "Religion, Social Networks, and Subjective Well-Being," *American Sociological Review* 75, no. 6 (2010): 914–33. Also see "Study Reveals 'Secret Ingredient' in Religion That Makes People Happier," American Sociological Association press release, December 7, 2012.

129 **Close friendships predict church attendance:** Tom Rath, *Vital Friends: The People You Can't Afford to Live Without* (New York: Gallup Press, 2006).

129 **"I had a very bad childhood":** E-mail and phone interviews with Charlotte Cook, February 24, 2011, and March 4, 2011.

131 **"Friendships provide feelings of worth":** Carolyn McNamara Barry and Stephanie D. Madsen, "Friends and Friendships in Emerging Adulthood," Changing Spirituality of Emerging Adults Project, http://changingsea.org/barry.htm.

131 **One reason may be that they don't feel:** J. Pulakos, "Young Adult Relationships: Siblings and Friends," *Journal of Psychology* 123 (1989): 237–44.

132 **Specifically, friends are better:** Phone interview with Simine Vazire, December 12, 2011.

132 **In fact, one study found that close friends tell fewer lies:** Bella M. DePaulo and Deborah A. Kashy, "Everyday Lies in Close and Casual Relationships," *Journal of Personality and Social Psychology* 74 (1998): 63–79.

133 **"Soliciting the information directly":** Phone interview with Simine Vazire, December 12, 2011.

133 **Making an effort to know your friends:** Charity A. Friesen and Laura K. Kammrath, "What It Pays to Know About a Close Other: The Value of If-Then Personality Knowledge in Close Relationships," *Psychological Science* 22, no. 5 (May 2011): 567–71.

134 **One study provides a particularly touching:** S. Schnall, K. D. Harber, J. K. Stefanucci, and D. R. Proffitt, "Social Support and the Perception of Geographical Slant," *Journal of Experimental Social Psychology* 44 (2008): 1246–55

134 **Tom Rath writes about a study:** Rath, *Vital Friends*.

135 **"There's a concept in ADHD":** Interview with David D. Nowell, March 9, 2011.

135 **Ian Anderson, a British man:** E-mail interview with Ian Anderson, March 2, 2011.

136 **Freya Harrison, Ph.D.:** F. Harrison, J. Sciberras, and R. James, "Strength of Social Tie Predicts Cooperative Investment in a Human Social Network," PLoS ONE, March 30, 2011, http://www.plosone.org/home.action.

137 **In their network analysis:** Christakis and Fowler, *Connected.*

138 **"A friend doesn't have to tell you":** Phone interview with James Fowler, December 8, 2011.

138 **For example, a new online experiment:** Carolyn Y. Johnson, "White Coat Notes: MIT Tests How Healthy Behaviors Spread in a Social Network," *Boston Globe*, December 1, 2011.

138 **Jeff Bell, a San Francisco radio personality:** Jeff Bell, "You've Got a Friend," Psychologytoday.com, July 23, 2010.

139 **Specifically, women feel less anxious:** S. L. Brown, B. L. Fredrickson, M. Wirth, M. Poulin, E. Meier, E. Heaphy, M. Cohen, and O. Schultheiss, "Social Closeness Increases Salivary Progesterone in Humans," *Hormones and Behavior* 56 (2009): 108–11.

139 **Laughing with friends can increase:** "Laughing with Others Eases Pain, Study Says," Foxnews.com, September 14, 2011.

139 **merely imagining a friend:** Phone interview with Julianne Holt-Lunstad, December 7, 2011.

139 **Older adults concerned about memory:** Karen A. Ertel, M. Maria Glymour, and Lisa F. Berkman, "Effects of Social Integration on Preserving Memory Function in a Nationally Representative U.S. Elderly Population," *American Journal of Public Health* 98, no. 7 (July 2008): 1215–20.

139 **You may have read about:** Candyce H. Kroenke, Laura D. Kubzansky, Eva S. Schernhammer, Michelle D. Holmes, and Ichiro Kawachi, "Social Networks, Social Support, and Survival After Breast Cancer Diagnosis," *Journal of Clinical Oncology* 24, no. 7 (March 1, 2006): 1105–11. Also see Tara Parker Pope, "What Are Friends For? A Longer Life," *New York Times,* April 20, 2009.

139 **Having a romantic attachment:** K. Orth-Gomer, A. Rosengren, and L. Wilhelmsen, "Lack of Social Support and Incidence of Coronary Heart Disease in Middle-Aged Swedish Men," *Psychosomatic Medicine* 55, no. 1 (1993): 37–43. Also see Pope, "What Are Friends For? A Longer Life."

140 **Julianne Holt-Lunstad, Ph.D., professor of psychology:** J. Holt-Lunstad, T. B. Smith, and J. B. Layton, "Social Relationships and Mortality Risk: A Meta-Analytic Review," *Public Library of Science Medicine* 7, no. 7 (2010).

140 **There is not just one answer:** Phone interview with Julianne Holt-Lunstad, December 7, 2011.

140 **Friends, in fact, are particularly:** Christakis and Fowler, *Connected.*

140 **That said, young adults often:** Interview with Meliksah Demir, October 12, 2010.

141 **Shelly, a thirty-year-old Canadian:** E-mail interview with Shelly, whose name has been changed, December 29, 2011.

141 **Richard Slatcher, Ph.D., an assistant professor of psychology:** Bonnie Rochman, "Why Hanging Out with Couple-Friends Enhances Romance," *Time,* February 16, 2011.

142 **As sociologist Mark Granovetter:** Mark Granovetter, The Strength of Weak Ties," *American Journal of Sociology* 78, no. 6 (May 1973): 1360–80.

142 **"If you are rich, you can attract":** Christakis and Fowler, *Connected.*

142 **A little over a decade ago:** Jan Yager, *Friendshifts: The Power of Friendship and How It Shapes Our Lives* (Stamford, Conn.: Hannacroix Creek Books, 1999).

142 **Tom Rath, of the Gallup Organization:** Rath, *Vital Friends.*

143 **In a finding that ties together:** Arie Shirom, Sharon Toker, Yasmin Alkaly, Orit Jacobson, and Ran Balicer, "Work-Based Predictors of Mortality: A 20-Year Follow-up of Healthy Employees," *Health Psychology* 30, no. 3 (May 2011): 268–75.

144 **Cross-race bonds**: Rodolfo Mendoza-Denton, "This Holiday, a Toast to Cross-Race Friendship," Psychologytoday.com, November 23, 2010.

144 **To answer the question:** Interview with James Vela-McConnell, April 11, 2011.

144 **Dalton Conley, Ph.D., chronicler:** Interview with Dalton Conley, April 19, 2011.

145 **Charles Duhigg, *New York Times* business:** Interview with Charles Duhigg, December 7, 2011.

145 **"such altruism often relies":** Daniel J. Hruschka, *Friendship: Development, Ecology, and Evolution of a Social Relationship* (Berkeley: University of California Press, 2010).

CHAPTER 6

147 **Shane Shaps still remembers:** Phone interview with Shane Shaps, February 6, 2012. (Claudia is a pseudonym for Shane's former friend.)

149 **One case study from the art world:** Angelique Chrisafis, "Art Historians Claim van Gogh's Ear 'Cut Off by Gauguin,'" guardian.co.uk, May 4, 2009.

150 **Gretchen Rubin, author of a blog and book:** Gretchen Rubin, "Are You Drifting?," www.happiness-project.com, July 22, 2009.

152 **"Over a quarter say that friends":** Mark Vernon, *The Meaning of Friendship* (New York: Palgrave Macmillan, 2010).

153 **Julianne Holt-Lunstad became interested in:** Phone interview with Julianne Holt-Lunstad, December 7, 2011.

154 **"But when we looked at the actual dates":** Ibid.

154 **This study also confirms:** Briahna Bigelow Bushman and Julianne Holt-Lunstad, "Understanding Social Relationship Maintenance Among Friends: Why We Don't End Those Frustrating Friendships," *Journal of Social and Clinical Psychology* 28, no. 6 (June 2009): 749–78.

154 **Jessica Chiang, a graduate student:** Jessica J. Chiang, Naomi L. Eisenberger, Teresa E. Seeman, and Shelley E. Taylor, "Negative and Competitive Social Interactions Are Related to Heightened Proinflammatory Cytokine Activity," *Proceedings of the National Academy of Sciences* (January 23, 2012).

154 **"Inflammation is a healthy response":** Phone interview with Jessica Chiang, February 7, 2012.

155 **A Western Michigan University study:** Ewa Urban, "Competition and Interpersonal Conflict in Same-Sex Platonic Friendships," *Hilltop Review* 1, no. 1 (2005).

155 **Our close buddies, in fact:** R. Weylin Sternglanz and Bella M. DePaulo, "Reading Nonverbal Cues to Emotions: The Advantages and Liabilities of Relationship Closeness," *Journal of Nonverbal Behavior* 28 (2004): 245–66.

156 **One honesty killer is the need:** Jan Yager, *When Friendship Hurts: How to Deal with Friends Who Betray, Abandon, or Wound You* (New York: Fireside, 2002).

157 **Mark Vernon sees rampant dishonesty:** Vernon, *The Meaning of Friendship*.

158 **Sixty-eight percent of those:** Yager, *When Friendship Hurts*.

159 **Shedding friends naturally:** Alex Williams, "It's Not Me, It's You," *New York Times*, January 28, 2012.

159 **"Over and over again":** Yager, *When Friendship Hurts*.

159 **Parting ways with a buddy:** Phone interview with Susan Shapiro Barash, February 6, 2012.

162 **Last year a nineteen-year-old:** "Latino Blotter: Latina Scalped During Girl on Girl Fight," Hispanicallyspeakingnews.com, November 3, 2011.

162 **Kelly Valen conducted a survey:** Kelly Valen, *The Twisted Sisterhood: Unraveling the Dark Legacy of Female Friendships* (New York: Ballantine Books, 2010).

162 **This study has a few problems:** Pamela Paul, "Mean Girls and Bad Mommies," *New York Times,* November 12, 2010.

163 **Yet Peter DeScioli, Ph.D., and Robert Kurzban, Ph.D.:** P. DeScioli and R. Kurzban, "The Company You Keep: Friendship Decisions from a Functional Perspective," in *Social Judgment and Decision Making*, ed. J. I. Krueger (New York: Psychology Press, 2011).

164 **Barash, who is a professor of gender studies:** Phone interview with Susan Shapiro Barash, February 6, 2012.

164 **Women often want what:** Ibid.

165 **Clinical psychologist Terri Apter, Ph.D.:** Phone interview with Terri Apter, February 7, 2012.

166 **An in-depth study by Robin Moremen, Ph.D.:** Robin D. Moremen, "The Downside of Friendship: Sources of Strain in Older Women's Friendship," *Journal of Women & Aging* 20, nos. 1 and 2 (2008): 169–87.

166 **Moremen is particularly concerned:** Ibid.

167 **Geoffrey Greif has found:** Jeffrey Zaslow, "Friendship for Guys (No Tears!)," *Wall Street Journal*, April 7, 2010.

167 **A Cornell University and University of Chicago study:** Benjamin Cornwell and Edward O. Laumann, "Network Position and Sexual Dysfunction: Implications of Partner Betweenness for Men," *American Journal of Sociology* 117 (2011): 172–208.

168 **The network research conducted:** Nicholas A. Christakis and James H. Fowler, *Connected: The Surprising Power of Our Social Networks and How They Shape Our Lives* (New York: Little, Brown and Company, 2009).

168 **In a follow-up study:** J. Niels Rosenquist, Joanne Murabito, James H. Fowler, and Nicholas A. Christakis, "The Spread of Alcohol Consumption Behavior in a Large Social Network," *Annals of Internal Medicine* 152, no. 7 (April 6, 2010): 426–33.

168 **a Scottish study found that:** Nick Collins, "Middle-Aged Drive to Avoid Peer Pressure of Drinking," *Telegraph*, December 12, 2011.

168 **For those battling drug addictions:** Phone interview with Carl Latkin, December 8, 2011.

169 **Raj Rajaratnam, a former hedge fund manager:** Peter Lattman and Azam Ahmed, "A Circle of Tipsters Who Shared Illicit Secrets," *New York Times*, May 11, 2011.

170 **In 2001, when they were both:** Maria Glod, "Daughter Gets 48 Years in Slaying of Her Father, Siblings Say No Sentence Could Bring Justice," *Washington Post*, February 11, 2003.

170 **It sounds like a benign, if nerdy:** Ibid.

171 **In contemplating the growing class divide:** David Brooks, "The Great Divorce," *New York Times*, January 30, 2012.

172 **"Modern friendship is supposed":** Interview with Dalton Conley, April 19, 2011.

173 **"a sort of secession":** C. S. Lewis, *The Four Loves* (Orlando, Fla.: Harcourt, 1991). Also see Matt Kaufman, "C. S. Lewis on 'The Dangers of Friendship,'" The Boundless Line (boundlessline.org), August 23, 2010.

173 **We know that preserving:** Irving L. Janis, *Victims of Groupthink: A Psychological Study of Foreign-Policy Decisions and Fiascoes* (New York: Houghton Mifflin, 1972).

173 **The greater the pack of friends:** Daniel Goleman, "As Bias Crime Seems to Rise, Scientists Study Roots of Racism," *New York Times*, May 29, 1990.

173 **Daniel Hruschka describes how:** Daniel J. Hruschka, *Friendship: Development, Ecology, and Evolution of a Social Relationship* (Berkeley: University of California Press, 2010).

174 **You'd think students at:** Angela J. Bahns, Kate M. Pickett, and Christian S. Crandall, "Social Ecology of Similiarity: Big Schools, Small Schools and Social Relationships," *Group Processes and Intergroup Relations* 15 (January 2012): 119–31.

174 **"They took a more fine-grained":** Phone interview with Angela Bahns, February 7, 2012.

175 **"My guess is that the higher":** Ibid.

175 **Since the 1970s, Cuban dictator:** Angel Esteban and Stephanie Panichelli-Batalla, *Fidel & Gabo: A Portrait of the Legendary Friendship Between Fidel Castro and Gabriel García Márquez* (New York: Pegasus Books, 2009).

176 **"The first phase of this relationship":** Phone interview with Stephanie Panichelli-Batalla, February 9, 2012.

176 **Social neuroscientist John Cacioppo, Ph.D.:** John T. Cacioppo and William Patrick, *Loneliness: Human Nature and the Need for Social Connection* (New York: W. W. Norton & Company, 2008).

CHAPTER 7

178 **Strange as it sounds:** Phone interview with Toni Bernhard, March 9, 2012.

182 **But sociologist Keith Hampton, Ph.D.:** Phone interview with Keith Hampton, March 20, 2012.

182 **But during that heralded:** Hua Wang and Barry Wellman, "Social Connectivity in America, Changes in Adult Friendship Network Size from 2002 to 2007," *American Behavioral Scientist* 53, no. 8 (April 2010): 1148–69.

182 **While it's too early to predict:** Phone interview with Keith Hampton, March 20, 2012.

183 **It's 150—that same old:** Chris Taylor, "Social Networking 'Utopia' Isn't Coming," CNN.com, June 27, 2011.

184 **As of late 2011:** Mary Madden and Kathryn Zickuhr, "65% of Online Adults Use Social Networking Sites," August 26, 2011, Pew Research Center's Internet & American Life Project, http://pewinternet.org/Reports/2011/Social-Networking-Sites.aspx.

184 **Internet-using Americans are more likely:** Lee Rainie, Kristen Purcell, and Aaron Smith, "The Social Side of the Internet," January 18, 2011,

Pew Research Center's Internet & American Life Project, http://pewinternet.org/Reports/2011/The-Social-Side-of-the-Internet/Summary.aspx.

184 **University of Texas, Austin, researchers:** S. Craig Watkins and H. Erin Lee, "Got Facebook? Investigating What's Social About Social Media," November 18, 2010, theyoungandthedigital.com.

185 **Keith Hampton uncovered the existence of:** Keith Hampton, Lauren Sessions Goulet, Cameron Marlow, and Lee Rainie, "Why Most Facebook Users Get More than They Give," February 3, 2012, Pew Research Center's Internet & American Life Project, http://www.pewinternet.org/Reports/2012/Facebook-users.aspx.

185 **Hampton points out that:** Phone interview with Keith Hampton, March 20, 2012.

185 **Sixty-three percent of users:** Lee Rainie, Amanda Lenhart, and Aaron Smith, "The Tone of Life on Social Networking Sites," February 9, 2012, Pew Research Center's Internet & American Life Project, http://pewinternet.org/Reports/2012/Social-networking-climate.aspx.

185 **Pamela Rutledge, Ph.D., the director:** Phone interview with Pamela Rutledge, March 15, 2012.

187 **Clay Shirky, Internet and society guru:** Clay Shirky, *Cognitive Surplus: How Technology Makes Consumers into Collaborators* (New York: Penguin, 2011).

188 **In an essay where he argues:** William Deresiewicz, "Faux Friendship," *Chronicle of Higher Education*, December 6, 2009, http://chronicle.com/article/Faux-Friendship/49308.

188 **Facebook critics will light up:** Ibid.

190 **Philosopher Mark Vernon calls:** Mark Vernon, *The Meaning of Friendship* (New York: Palgrave Macmillan, 2010).

191 **"The essential trouble with e-mail":** E-mail interview with Henry Alford, April 3, 2012.

191 **Andrea Bonior, Ph.D., clinical psychologist:** Phone interview with Andrea Bonior, March 9, 2012.

192 **Bonior has seen shy clients:** Ibid.

192 **A study from the University of Waterloo:** Amanda L. Forest and Joanne V. Wood, "When Social Networking Is Not Working: Individuals with Low Self-Esteem Recognize but Do Not Reap the Benefits of Self-Disclosure on Facebook," *Psychological Science* 23, no. 3 (March 2012): 295–303.

192 **Incidentally, a recent study found:** Christopher J. Carpenter, "Narcissism on Facebook: Self-Promotional and Anti-Social Behavior," *Personality and Individual Differences* 52, no. 4 (March 2012): 482–86.

193 **Sociologists Hui-Tzu Grace Chou, Ph.D., and Nicholas Edge:**

Hui-Tzu Grace Chou and Nicholas Edge, "'They Are Happier and Having Better Lives than I Am': The Impact of Using Facebook on Perceptions of Others' Lives," *Cyberpsychology, Behavior, and Social Networking* 15, no. 2 (February 2012): 117–21.

193 **Bonior sees envy crop up:** Phone interview with Andrea Bonior, March 9, 2012.

195 **"probably the most connected":** Vernon, *The Meaning of Friendship*.

195 **An Italian team recorded:** Maurizio Mauri, Pietro Cipresso, Anna Balgera, Marco Villamira, and Giuseppe Riva, "Why Is Facebook So Successful? Psychophysiological Measures Describe a Core Flow State While Using Facebook," *Cyberpsychology, Behavior, and Social Networking* 14, no. 12 (December 2011): 723–31.

196 **Larry Rosen, Ph.D., the author:** Phone interview with Larry Rosen, March 12, 2012.

196 **Among the plugged-in:** Interview with Jacob, whose name has been changed, February 25, 2012.

197 **Perhaps the scariest dispatch:** Miller McPherson, Lynn Smith-Lovin, and Matthew E. Brashears, "Social Isolation in America: Changes in Core Discussion Networks over Two Decades," *American Sociological Review* 71, no. 3 (June 2006): 353–75.

197 **When Hua Wang and Barry Wellman, Ph.D., compared:** Hua Wang and Barry Wellman, "Social Connectivity in America, Changes in Adult Friendship Network Size from 2002 to 2007," *American Behavioral Scientist* 53, no. 8 (April 2010): 1148–69.

197 **Keith Hampton and colleagues discovered:** Keith Hampton, Lauren Sessions, and Eun Ja Her, "Core Networks, Social Isolation, and New Media: Internet and Mobile Phone Use, Network Size and Diversity," *Information, Communication and Society* 14, no. 1 (2011): 130–55.

198 **For example, a study of middle-aged:** Cynthia M. H. Bane, Marilyn Cornish, Nicole Erspamer, and Lia Kampman, "Self-Disclosure Through Weblogs and Perceptions of Online and 'Real-life' Friendships Among Female Bloggers," *Cyberpsychology, Behavior, and Social Networking* 13, no. 2 (April 2010): 131–39.

198 **As researcher Lijun Tang:** Lijun Tang, "Development of Online Friendship in Different Social Spaces," *Information, Communication and Society* 13, no. 4 (2010): 615–33.

198 **"On social media sites, people":** Phone interview with Larry Rosen, March 12, 2012.

198 **Rosen has a positive view:** Ibid.

199 **"has as many cheerleaders":** Vernon, *The Meaning of Friendship*.

199 **Pete Beatty, age thirty:** Phone interview with Pete Beatty, March 9, 2012.

201 **"The real question is":** Phone interview with Pamela Rutledge, March 15, 2012.

202 **Arikia Millikan, now twenty-five:** Phone interview with Arikia Millikan, March 13, 2012.

204 **Stanford University surveyed preteen girls:** Roy Pea, Clifford Nass, Lyn Meheula, Marcus Rance, Aman Kumar, Holden Bamford, Matthew Nass, Aneesh Simha, Benjamin Stillerman, Steven Yang, Michael Zhou, "Media Use, Face-to-Face Communication, Media Multitasking, and Social Well-Being Among 8- to 12-Year-Old Girls." *Developmental Psychology* 48, no. 2 (March 2012): 327–36.

205 **Larry Rosen has studied:** Phone interview with Larry Rosen, March 12, 2012.

205 **Danah Boyd, Ph.D., a fellow:** Pamela Paul, "Cracking Teenagers' Online Codes," *New York Times*, January 20, 2012.

206 **One preference that apparently doesn't:** Eric Smalley, "Are We Immune to Viral Marketing?," Wired.com, December 19, 2011.

207 **Irfan Kamal, the senior vice president:** Ibid.

207 **Paul Adams, a researcher:** Video presentation, Paul Adams, "How Our Social Circles Influence What We Do, Where We Go, and How We Decide," UX Week, San Francisco, August 23–26, 2011.

208 **Tina Rosenberg, a writer and expert:** Tina Rosenberg, "On Gay Rights, Moving Real-Life Friends to Action," *New York Times*, July 7, 2011.

209 **At a recent tech conference:** Jenna Wortham, "New Apps Connect to Friends Nearby," *New York Times*, March 8, 2012.

209 **Michelle Norgan, a cofounder:** Ibid.

CHAPTER 8

210 **Renee Young met her best friend:** E-mail interview with Renee Young, February 23, 2011.

211 **"obligatory gregarious":** John T. Cacioppo and William Patrick, *Loneliness: Human Nature and the Need for Social Connection* (New York: W. W. Norton & Company, 2008).

213 **Mark Vernon worries that thinking:** Mark Vernon, *The Meaning of Friendship* (New York: Palgrave Macmillan, 2010).

215 **Since May 2007:** Phone interview with Irene Levine, March 30, 2012.

216 **"Each of us inherits from our parents":** Cacioppo and Patrick, *Loneliness*.

216 **"If you have one friend":** Phone interview with John Cacioppo, April 3, 2012.

217 **Tom Rath, of the Gallup Organization:** Tom Rath, *Vital Friends: The People You Can't Afford to Live Without* (New York: Gallup Press, 2006).

217 **It's also illuminating:** Ibid.

218 **Does scheduling time together:** Irene S. Levine, *Best Friends Forever: Surviving a Breakup with Your Best Friend* (New York: Overlook Press, 2009).

219 **"It's very difficult to draw a line":** Phone interview with Terri Apter, February 7, 2012.

219 **Andrea Bonior suggests "slowly backing away":** Phone interview with Andrea Bonior, March 9, 2012.

220 **Of all those letters on friend dilemmas:** Phone interview with Irene Levine, March 30, 2012.

221 **Unfortunately, though, its effects:** Cacioppo and Patrick, *Loneliness.*

221 **"Most neuroscientists now agree":** Ibid.

222 **Relatedly, a study by Sigal Barsade, Ph.D.:** Phyllis Korkki, "Building a Bridge to a Lonely Colleague," *New York Times,* January 28, 2012.

222 **"Feeling lonely does not mean":** Cacioppo and Patrick, *Loneliness.*

222 **Introverts get pegged as awkward:** Phone interview with John Cacioppo, April 3, 2012.

222 **If a person said he was lonely:** John T. Cacioppo, James H. Fowler, and Nicholas A. Christakis, "Alone in the Crowd: The Structure and Spread of Loneliness in a Large Social Network," *Journal of Personality and Social Psychology* 97, no. 6 (December 2009): 977–91.

223 **"There is a mechanism":** Phone interview with John Cacioppo, April 3, 2012.

223 **This mechanism is turbocharged:** Ibid.

223 **But his ultimate message:** Cacioppo and William Patrick, *Loneliness.*

224 **"People tell me all the time":** Phone interview with Andrea Bonior, March 9, 2012.

224 **That is precisely what the journalist:** Phone interview with Rachel Bertsche, December 12, 2011. A portion of this interview was published as Carlin Flora, "Have You Ever Friend-Dated?," Psychologytoday.com, December 21, 2011.

225 **"She writes about the greater trust":** Cacioppo and Patrick, *Loneliness.*

225 **"If you live in a neighborhood":** Phone interview with John Cacioppo, April 3, 2012.

226 **"I was very isolated":** Interview with Joseph (last name withheld), April 2, 2012.

226 **One study found that lonely people:** Xinyue Zhou, Constantine Sedikides, Tim Wildschut, and Ding-Guo Gao, "Counteracting Loneliness: On the Restorative Function of Nostalgia," *Psychological Science* 19, no. 10 (2008): 1023–29.

227 **Some symptoms of social anxiety:** Levine, *Best Friends Forever.*

227 **"The therapist will give patients":** Phone interview with Andrea Bonior, March 9, 2012.

227 **As a high-functioning woman:** E-mail interview with Lynne Soraya, April 2, 2012.

229 **"In my twenties and thirties":** Interview with Jacob, whose name has been changed, February 25, 2012.

230 **"I loved it when a friend whose":** E-mail interview with Henry Alford, April 3, 2012.

231 **Daniel Hruschka reviewed studies:** Daniel J. Hruschka, *Friendship: Development, Ecology, and Evolution of a Relationship* (Berkeley: University of California Press, 2010).

231 **"Because close friends are seen":** Interview with Dalton Conley, April 19, 2011.

231 **Now that Jacob is planning:** Interview with Jacob, whose name has been changed, February 25, 2012.

231 **"It was appealing not just for":** Phone interview with Terri Apter, February 7, 2012.

232 **Those who dispense friend advice:** Hruschka, *Friendship.*

232 **Jacob, who earns a high salary:** Interview with Jacob, whose name has been changed, February 25, 2012.

232 **"When men are doing badly":** Interview with an ad executive who didn't want his name used, December 15, 2011.

233 **"Be careful about saying things":** Jan Yager, *When Friendship Hurts: How to Deal with Friends Who Betray, Abandon, or Wound You* (New York: Fireside, 2002).

233 **"It's more awkward for me":** Interview with Dalton Conley, April 19, 2011.

234 **Hruschka writes of a study:** Hruschka, *Friendship.*

234 **Consider these gems from:** Tom Sienkewicz, "Cicero's Laelius de Amicitia ('Laelius on Friendship'): A Summary," department.monm.edu/classics/courses/clas210/coursedocuments/cicero_on_friendship_a_summary.htm.

235 **Bella DePaulo has documented and critiqued:** Bella DePaulo, *Single with Attitude: Not Your Typical Take on Health and Happiness, Love and Money, Marriage and Friendship* (Seattle: CreateSpace, 2009).

INDEX